PROGRESS IN

Nucleic Acid Research and Molecular Biology

Volume 77

PROGRESS IN
Nucleic Acid Research and Molecular Biology

edited by

KIVIE MOLDAVE

Department of Molecular Biology and Biochemistry
University of California, Irvine
Irvine, California

Volume 77

AMSTERDAM • BOSTON • HEIDELBERG • LONDON
NEW YORK • OXFORD • PARIS • SAN DIEGO
SAN FRANCISCO • SINGAPORE • SYDNEY • TOKYO

Academic Press is an imprint of Elsevier

Elsevier Academic Press
525 B Street, Suite 1900, San Diego, California 92101-4495, USA
84 Theobald's Road, London WC1X 8RR, UK

This book is printed on acid-free paper. ∞

Copyright © 2004, Elsevier Inc. All Rights Reserved.

No part of this publication may be reproduced or transmitted in any form or by any means, electronic or mechanical, including photocopy, recording, or any information storage and retrieval system, without permission in writing from the Publisher.

The appearance of the code at the bottom of the first page of a chapter in this book indicates the Publisher's consent that copies of the chapter may be made for personal or internal use of specific clients. This consent is given on the condition, however, that the copier pay the stated per copy fee through the Copyright Clearance Center, Inc. (www.copyright.com), for copying beyond that permitted by Sections 107 or 108 of the U.S. Copyright Law. This consent does not extend to other kinds of copying, such as copying for general distribution, for advertising or promotional purposes, for creating new collective works, or for resale. Copy fees for pre-2004 chapters are as shown on the title pages. If no fee code appears on the title page, the copy fee is the same as for current chapters.
0079-6603/2004 $35.00

Permissions may be sought directly from Elsevier's Science & Technology Rights Department in Oxford, UK: phone: (+44) 1865 843830, fax: (+44) 1865 853333, e-mail: permissions@elsevier.com.uk. You may also complete your request on-line via the Elsevier homepage (http://elsevier.com), by selecting "Customer Support" and then "Obtaining Permissions."

For all information on all Academic Press Publications
visit our Web site at www.academicpress.com

ISBN: 0-12-540077-2

PRINTED IN THE UNITED STATES OF AMERICA
04 05 06 07 08   9 8 7 6 5 4 3 2 1

Contents

SOME ARTICLES PLANNED FOR FUTURE VOLUMES ix

Nuclear Receptor-Mediated Transactivation Through Interaction with Sp Proteins 1

Stephen Safe and Kyounghyun Kim

I. Sp1 Protein 2
II. Sp/KLF Family of Proteins 3
III. Cooperative Sp1 and Related Sp/KLF Protein Interactions with Other Transcription Factors 4
IV. Interactions of Sp Family Proteins with Estrogen Receptors 9
V. Interaction of Sp Family Proteins with Other Steroid Hormone Receptors and NRs 20
VI. Summary 23
References 24

Site-Specific DNA Damage Recognition by Enzyme-Induced Base Flipping 37

James T. Stivers

I. Why Do Enzymes Flip Bases? 39
II. Nonenzymatic Base Pair Breathing 41
III. Enzymatic Base Flipping 43
IV. New Experimental Approaches 54
V. Future Directions 60
References 61

Bacteriophage T2Dam and T4Dam DNA-[N6-adenine]-methyltransferases 67

Stanley Hattman and Ernst G. Malygin

I. Historical Background 68
II. Binding Properties of T4Dam 73
III. Kinetic Properties of T4Dam Methylation of Substrate Duplexes Containing Native or Altered Recognition Sites 84

IV. Structure of T4Dam	112
V. Concluding Comments	118
References	120

Site-Specific Recombination and Partitioning Systems in the Stable High Copy Propagation of the 2-Micron Yeast Plasmid 127

Makkuni Jayaram, Shwetal Mehta, Dina Uzri, Yuri Voziyanov, and Soundarapandian Velmurugan

I. The Flp Site-Specific Recombination System	130
II. The Shared Active Site of Flp: DNA Cleavage in *trans*	135
III. The Geometry and Topology of Flp Recombination	142
IV. The 2-Micron Circle Partitioning System	151
V. Cohesin Disassembly is a Prerequisite for Separation of Plasmid Clusters	161
VI. Summary and Perspectives	164
References	167

Did an Early Version of the Eukaryal Replisome Enable the Emergence of Chromatin? 173

Gabriel Kaufmann and Tamar Nethanel

I. Introduction	173
II. Deep-Rooted Features of Two Replisome Types	174
III. Coordinated Syntheses of the Opposite DNA Strands	187
IV. Possible Implications of a Symmetrical Eukaryal Replication Fork	196
References	199

Initiation and Elongation Factors in Mammalian Mitochondrial Protein Biosynthesis. 211

Linda L. Spremulli, Angie Coursey, Tomas Navratil, and Senyene Eyo Hunter

I. Mammalian Mitochondrial Protein Synthesis	211
II. Initiation of Protein Biosynthesis in Mammalian Mitochondria	214
III. Elongation Factors in Mammalian Mitochondria	231
IV. Conclusions	252
References	253

Cyclin Dependent Kinase 11 in RNA Transcription and Splicing 263

Janeen H. Trembley, Pascal Loyer, Dongli Hu, Tongyuan Li, Jose Grenet, Jill M. Lahti, and Vincent J. Kidd

I. Introduction	263
II. CDK11 Gene Structure and Expression	264
III. CDK11 Protein Isoforms and Regulation	270
IV. CDK Regulation of RNA Transcription	275
V. CDK11^{p110} Protein Complexes and Biological Function	278
VI. Alterations in CDK11^{p110} Expression in Human Disease	282
References	283

The Eukaryotic Ccr4-Not Complex: A Regulatory Platform Integrating mRNA Metabolism with Cellular Signaling Pathways? 289

Martine A. Collart and H. Th. Marc Timmers

I. Introduction	290
II. The Ccr4-Not Complex: Conserved in Composition and Organization	292
III. Interaction of the Core Ccr4-Not Complex with Additional Proteins in Larger Structures	298
IV. Role of the Yeast Ccr4-Not Complex in mRNA Metabolism	305
V. Role of the Ccr4-Not Complex in Ubiquitylation	312
VI. Role of the Ccr4-Not Complex in Protein Modification	314
VII. The Ccr4-Not Complex as a Regulatory Platform that Senses Glucose Levels and Stress	314
VIII. Perspectives	316
References	317

Signaling Repression of Transcription by RNA Polymerase III in Yeast 323

Ian M. Willis, Neelam Desai, and Rajendra Upadhya

I. Introduction	323
II. Coupling Plasma Membrane Expansion to Transcriptional Regulation of Ribosomal Components and tRNA	326
III. Nutrients, Starvation, and TOR Signaling	335
IV. Repression of Pol III Transcription in Response to DNA Damage	341

V. Maf1 and its Essential Role in Pol III Transcriptional Repression 342
VI. Future Studies 346
   References 348

Index 355

Some Articles Planned for Future Volumes

Tandem CCCH Zinc Finger Proteins in the Regulation of mRNA Turnover
  PERRY BLACKSHEAR

Initiation and Recombination: Early and Late Events in the Replication of Herpes Simplex Virus
  PAUL E. BOEHMER

Molecular Regulation, Evolutionary and Functional Adaptations Associated with C to U Editing of Mammalian Apolipoprotein B mRNA
  NICHOLAS O. DAVIDSON, SHRIKANT ANANT, AND VALERIE BLANC

Conformational Polymorphism of d(A-G)n and Related Oligonucleotide Sequences
  JACQUES FRESCO AND NINA G. DOLINNAYA

DNA-Protein Interactions Involved in the Initiation and Termination of Plasmid Rolling Circle Replication
  SALEEM A. KAHN, T.-L. CHANG, M. G. KRAMER, AND M. ESPINOSA

FGF3: A Gene with a Finely Tuned Spatiotemporal Pattern of Expression During Development
  CHRISTIAN LAVIALLE

Specificity and Diversity In DNA Recognition by *E. Coli* Cyclic AMP Receptor Protein
  JAMES C. LEE

Steroid Signaling in Procaryotes
  EDMOND MASER

Oxygen Sensing and Oxygen-regulated Gene Expression in Yeast
  ROBERT O. POYTON

Ribonucleases in Cancer Chemotherapy
  ROBERT T. RAINES, P. A. LELAND, M. C. HERBERT, AND K. E. STANISZEWSKI

Broad Specificity of Serine/Arginine (SR)-rich Proteins Involved in the Regulation of Alternative Splicing of Premessnger RNA
  JAMES STEVENIN, CYRIL BOURGEOIS, AND FABRICE LEJUNE

DNA Double-Strand Break Repair in Eukaryotic Cells
  PATRICK SUNG, LUMIR KREJCI, AND ALAN TOMKINSON

FOXO Forkhead Transcription Factors in Insulin and Growth Factor Action
  TERRY UNTERMAN, SHAODONG GUO, AND XIAOHUI ZHANG

Nuclear Receptor-Mediated Transactivation Through Interaction with Sp Proteins[1]

STEPHEN SAFE AND
KYOUNGHYUN KIM

*Department of Veterinary Physiology and
Pharmacology, Texas A&M University,
College Station, TX 77843-4466*

I. Sp1 Protein	2
II. Sp/KLF Family of Proteins	3
III. Cooperative Sp1 and Related Sp/KLF Protein Interactions with Other Transcription Factors	4
A. Promoter DNA-Dependent Sp/KLF-Protein Interactions	4
B. DNA-Independent Interactions of Transcription Factors with Sp1	7
IV. Interactions of Sp Family Proteins with Estrogen Receptors	9
A. DNA-Dependent Estrogen Receptor-Sp1 Interactions	9
B. ERα/Sp1 Activation of GC-Rich Promoters	12
C. Comparative Activation of GC-Rich Promoters by ERα/Sp1 and ERβ/Sp1	16
D. Domains of ERα Required for Estrogen and Antiestrogen Activation of ERα/Sp1	16
E. ERα/Sp3 Interactions with GC-Rich Promoters	19
F. Coactivation of ERα/Sp-Dependent Transactivation	20
V. Interaction of Sp Family Proteins with Other Steroid Hormone Receptors and NRs	20
A. Progesterone Receptor (PR)/Sp1-Mediated Transactivation	20
B. Androgen Receptor (AR)/Sp1-Mediated Transactivation	21
C. Retinoid Receptor/Sp1-Mediated Transactivation	21
D. PPARγ/Sp1-Mediated Transactivation	22
E. Chicken Ovalbumin Upstream Promoter-Transcription Factor (COUP-TF)/Sp1-Mediated Transactivation	22
F. Steroidogenic Factor-1 (SF-1)/Sp1-Mediated Transactivation	22
VI. Summary	23
References	24

---

[1] Arnt, AhR nuclear translocator; AhR, aryl hydrocarbon receptor; CKB, creatine kinase B; DRE, dioxin response element; ER, estrogen receptor; EREs, estrogen responsive elements; NRs, nuclear receptors; PPARs, peroxisome proliferator-activated receptors; RAR, retinoid acid receptor; RXR, retinoid X receptor; Sp1, specificity protein 1; SF-1, steroidogenic factor-1; TBP, TATA-binding protein; TAFs, TATA-binding protein-associated factors; TGFβ, transforming growth factor β.

I. Sp1 Protein

Specificity protein 1 (Sp1) was first identified by cell fractionation procedures and shown to interact with GC and GT oligonucleotide sequences that are typically found in diverse viral and cellular gene promoters (1–6). The Sp1 gene was first cloned by Kadonaga and coworkers, and the various functional domains of Sp1 were determined in a series of *in vitro* and whole cell assays. A schematic representation of the functional domains of Sp1 is given in Fig. 1. These domains include an N-terminal A- and a central B-domain which contain S/T- and Q-rich regions that are important for transactivation (7–10). Within the C- and D-domains near the C-terminal region of Sp1, there are three highly characteristic zinc fingers required for sequence-specific DNA binding.

The role of Sp1 as a transcription factor has been extensively investigated, and it is evident that the highly diverse functions of Sp1 are dependent on multiple factors, including promoter and cell context [reviewed in (11–14)]. The identification of transactivation domains in Sp1 and other Sp transcription factors implies that transcriptional activation involves interaction with other nuclear coregulatory proteins. Sp1-dependent transactivation from GC-rich promoters with TATA boxes or initiator sites (TATA-less) involves complex interactions with several TATA-binding protein (TBP)-associated factors (TAFs) that form part of the transcriptional machinery (15–18). TFIID is a protein complex containing TBP and multiple TAFs and is required for transcription by most RNA polymerase II-dependent gene promoters. Sp1-dependent transactivation involves Sp1 protein interactions with TFIID and specific TAF components of this complex (19, 20). *Drosophila* TAF110 (dTAF-110) specifically interacts with the transactivation domains A and B of Sp1 (21, 22). hTAF130 is the human equivalent of dTAF110; however, these TAFs display limited sequence similarity, and interactions of Sp1 with hTAF130 have been extensively investigated. The A-domain of Sp1 makes multiple

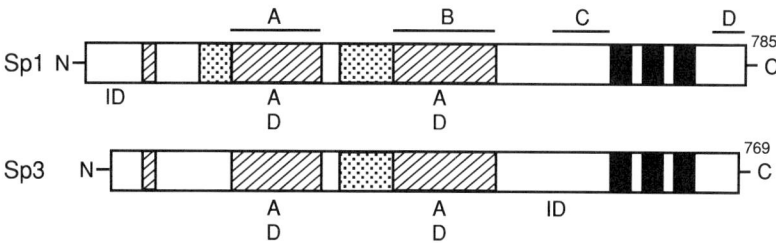

FIG. 1. Domains of Sp1 and Sp3 proteins. The multiple domains of Sp1 and Sp3 proteins include zinc fingers (■) involved in DNA binding, Q-rich (▨), and S/T-rich (▦) activation domains (AD) and inhibitory domains (ID).

contacts with glutamine-rich regions in hTAF130, whereas Sp1 B-domain interactions are localized to only one of the Q-regions (23, 24). Sp1 also binds other members of the basal transcription machinery including TBP (25) and human TAFII55, which interacts with the DNA binding domain of Sp1 (26). TAFII250 cooperatively activates Sp1 through direct interactions with retinoblastoma protein (27).

Sp1-dependent transactivation has also been linked to multiprotein nuclear complexes designated as cofactors required for Sp1 coactivation or CRSP (28–30). Biochemical fractionation and *in vitro* transcription assays identified a 700 kd complex containing at least nine subunits designated as CRSP200, CRSP150, CRSP130, CRSP100, CRSP85, CRSP77, CRSP70, CRSP34, and CRSP30. Studies in several laboratories have identified a similar series of pol II-interacting coactivator complexes, which contain many of the same common subunits (28–37). These complexes include vitamin D interacting proteins (DRIPs), thyroid hormone receptor associated proteins (TRAPs), activator-required cofactor, human mediator complexes (hMed), positive cofactor 2 (PC2), and SRB/MED containing cofactor complex (SMCC). The precise interactions of individual CRSP subunits with Sp1 have not been defined; however, the identical CRSP 200/DRIP 205/perosixome proliferator-activated receptor binding protein (PBP) identified in many of these complexes directly bind nuclear receptor and anchor mediator complex-receptor interactions (37). The association of mediators and TAFs with Sp1 illustrate the potential complexities of Sp1-dependent transactivation through cell context-dependent interactions with networks of coregulatory proteins.

II. Sp/KLF Family of Proteins

Sp1 interactions with GC-rich motifs and other "Sp1 binding sites" initially defined the prototypical sequence-specific binding of transcription factors; however, Sp1 is only one member of a family of zinc finger proteins that also bind these sites (11–14). Sp1, Sp2, Sp3, and Sp4 are structurally related Sp proteins that contain the characteristic three zinc fingers on the C-terminal region as well as activation domains and other common structural motifs. Sp5, Sp6, Sp7, Sp8, and Krüppel-like factors (KLFs) are also members of this Sp/KLF family, which are characterized by three zinc finger motifs in the C-terminal region, which interact with GC-rich elements. Compared to S1–Sp4, most of the remaining Sp/KLF family are lower molecular weight proteins that do not contain Q-rich activation domains. Sp/KLF transcription factors (with the exception of Sp2) bind common GC-rich gene promoter elements, and their potential role in gene regulation depends on multiple factors including tissue-/cell-specific expression patterns. In addition, Sp1 protein can also be modified by glycosylation and phosphorylation, and decreased glycosylation in glucose-deprived cells enhances proteasome-dependent degradation of Sp1 (38–46). In many cell lines, Sp3 decreases gene expression through an inhibitory domain containing a KEE amino acid triplet, and acetylation of the lysine residue contributes to the low transcriptional activity of Sp3 (47, 48).

Despite the multiplicity of Sp/KLF proteins and possible redundancies in their mechanisms of gene regulation, the knockout phenotypes of some Sp/KLF genes demonstrate their critical but distinct cellular functions. For example, Sp1 knockout mice exhibit multiple abnormalities, retarded development, and embryolethality on day 11 of gestation (49). Sp4 is primarily expressed in the brain and central nervous system, and two-thirds of Sp4 knockout mice die within a few days of birth. The surviving animals are reduced in size and the male mice are infertile (50–52). Sp5 knockout animals do not display a specific phenotype (53); bone formation is blocked in Sp7 knockout mice (54); and ossification is impaired in Sp3 knockout mice (55). Thus, it is evident that Sp proteins play a critical and varied role in cellular development and function, and the specificity of their function can be understood, in part, from studies on cell context-dependent regulation of genes through interactions of Sp proteins with GC-rich promoter elements.

III. Cooperative Sp1 and Related Sp/KLF Protein Interactions with Other Transcription Factors

A. Promoter DNA-Dependent Sp/KLF-Protein Interactions

Sp1-dependent transactivation from GC-rich promoters is not only coactivated by basal transcription factors but also by other nuclear proteins including DNA-bound transcription factors. Courey and coworkers (10) demonstrated synergistic activation of transcription by Sp1 through interactions with Sp1 variants containing activation domains. Moreover, enhanced transactivation through interactions between multiple Sp1 binding sites has also been observed using promoters with proximal or distal GC-boxes. GC-boxes and other Sp1 binding sites have been characterized in genes associated with cell proliferation, DNA synthesis, nucleotide metabolism, and interactions between Sp proteins are dependent on the gene promoter and cell context.

Research in this laboratory has investigated the effects of GC-rich sites on the basal activity of several gene promoter constructs in ER-positive MCF-7 and ZR-75 cells. Transfection of constructs containing only the proximal GC-rich motifs from the trifunctional carbamoylphosphate synthetase/aspartate carbamyltransferase/dihydroorotase (CAD) gene promoter (56) shows that successive 5' to 3' deletion of GC-rich sites 1–3 decreased basal activity from 100 to 30 to 10 in MCF-7 cells and 100 to 80 to 30 in ZR-75 cells (Fig. 2A). This suggested that contributions from site 1 were more important

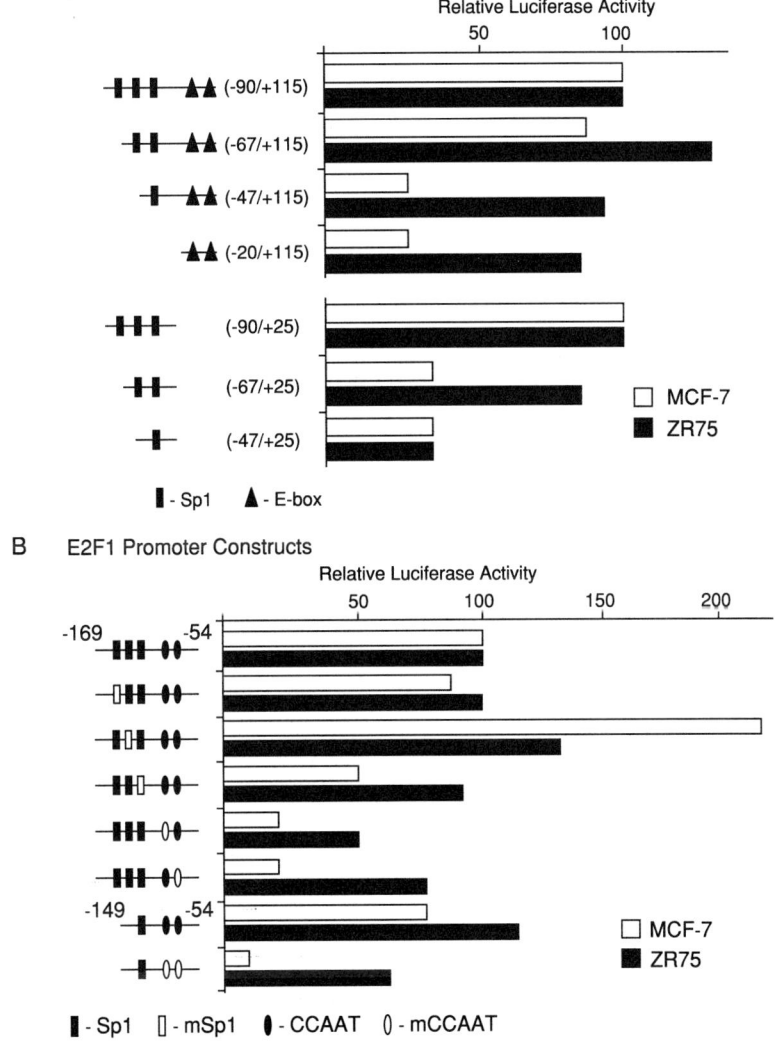

FIG. 2. Interactions between multiple GC-rich sites. Reporter gene activity in constructs containing CAD (A) and E2F1 (B) promoter inserts is dependent on specific GC-rich sites and cell context in ZR-75 and MCF-7 human breast cancer cells (56–58).

in MCF-7 than ZR-75 cells. Transfection of constructs containing the same three GC-rich sites plus two downstream E-box motifs gave higher (approximately 2-fold) basal activity in both cell lines, which is consistent with contributions of E-box binding proteins. Successive deletions of GC-boxes 1–3 using the E-box containing constructs showed that proteins binding GC-box 2 contributed significantly to basal activity in MCF-7 cells. In contrast, basal activity in ZR-75 cells was primarily due to the E-boxes, which bind USF1/2 in both cell lines and was independent of GC-boxes 1–3. The E2F1 gene promoter also contains three proximal GC-boxes and two downstream CCAAT motifs that bind NFYA/B (57, 58). Transfection of a series of deletion/mutant constructs containing E2F1 promoter inserts shows that basal activity of the constructs in ZR-75 cells is primarily due to cooperative effects of GC-boxes 1–3 and the 5'-CCAAT site (Fig. 2B). In contrast, both CCAAT motifs are of major importance for basal activity in MCF-7 cells, and results suggest that GC-box 2 binds "inhibitory" factors. This suggests that for some GC-rich promoters containing multiple GC-rich motifs, there may be preferential interaction of different Sp/KLF factors with specific GC-rich sites. This is currently being investigated.

Several studies report that modulation of the cyclin-dependent kinase inhibitor p21 is dependent on a series of 6 proximal GC-rich Sp/KLF sites (−154 to start site) (Fig. 3). The mouse p21 promoter is activated by both Sp1 and Sp3 in NIH3T3, HeLa and COS-1 cells, and increased activation of this construct by histone deacetylase inhibitors were associated with coactivation by p300 (59, 60). In contrast, activation of p21 through the proximal GC-rich elements in the human p21 promoter was primarily dependent on Sp3 and not Sp1 in MG63 cells (61–63), whereas, in other human cell lines and *Drosophila* cells, Sp1 was a more potent activator (64, 65). A recent paper showed that Krüppel-like factor 4 (KLF4) is involved in p53-mediated activation of p21

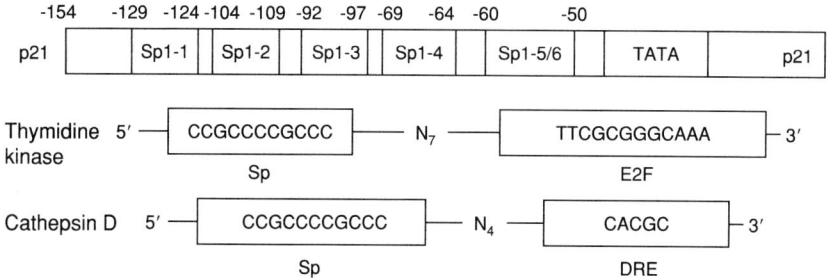

FIG. 3. GC-boxes in several gene promoters. The multiple (6) GC-boxes in the human p21 gene promoter and the GC boxes adjacent to E2F and core DRE motifs in the thymidine kinase and cathepsin D gene promoters, respectively, are indicated.

through interactions of KLF4 with a proximal GC-rich site in the human p21 promoter (66).

Sp1 and Sp protein-dependent interactions with other transcription factors and nuclear proteins play a key role in activation of multiple genes including p21. Moreover, interactions of these proteins can be both DNA-dependent and -independent. A classical example of a DNA-dependent interaction involves the cooperative activation of the thymidine kinase promoter by Sp1 and E2F1 (Fig. 3). Both E2F1 and Sp1 proteins physically interact in pulldown or coimmunoprecipitation assays. E2F1 interacts primarily with the C-terminal (aa 622–788) DNA binding domain of Sp1, whereas Sp1 binds to the N-terminal region (aa 1–112) of E2F1 (67, 68). The thymidine kinase gene promoter contains a GC-rich and an E2F1 site separated by 6 bp. Mutation of either site resulted in >90% loss of basal activity in Swiss 3T3 cells transfected with constructs containing the appropriate wild-type and mutant thymidine kinase promoter inserts. Moreover, as the distance between these sites was increased from 6 to 26 bp, there was a parallel decrease in transactivation, indicating that proximity of the two binding sites was required for maximal cooperative transactivation.

Similar interactions between Sp1 and the aryl hydrocarbon receptor (AhR) complex have been observed in MCF-7 cells transfected with a construct [pCD(−145/−119)] containing the −145 to −119 region of the cathepsin D gene promoter (69). This region of the promoter contains GC-rich and core dioxin response element (DRE) (GCGTG) motifs separated by 2 bp (Fig. 3). Mutation of one or both of these sites resulted in significant loss of activity, and increased separation of the GC-rich and DRE motif also decreased basal activity in transfected MCF-7 cells. The increased nucleotide spacing also affected AhR complex-Sp1 interactions in gel mobility shift assays, and it was shown by Kobayashi and coworkers (70) that AhR and AhR nuclear translocator (Arnt) coimmunoprecipitate with Sp1 and preferentially interact with the C-terminal domain of Sp1. These results illustrate DNA-dependent interactions of Sp1 with DNA-bound transcriptions and this has been observed with many other transcription factors.

B. DNA-Independent Interactions of Transcription Factors with Sp1

Interactions of Sp1 with other transcription factors indicate that transactivation can be the result of DNA-bound transcription factors or by activation of Sp1 through protein–protein interactions. Table I summarizes some of the reported interactions of transcription factors with Sp1 in which transactivation through Sp1-transcription factor interactions are DNA-dependent or -independent or both. The remarkable diversity of these interactions can be illustrated with studies on p21 regulation in various cell contexts. As has

TABLE I
Interaction of Sp1 with Other Nuclear Factors

Sp1-interacting factor	Interacting domain of Sp1	Reference
Ah receptor	C/D	(70)
Ah receptor nuclear translocator (Arnt)	C/D	(70)
GATA-1	C/D	(77)
GATA-2	C/D	(77)
GATA-3	C/D	(77)
NF-YA	A/B	(78–80)
Von-Hippel Lindau (VHL) tumor suppressor	C/D	(81, 82)
MyOD	C/D	(83)
Histone deacetylase 1 (HDAC1)	C/D	(84)
Promyelocytic leukemia protein (PML)	C/D	(85)
Helix-like transcription factor (HTLF)	C/D	(86)
E2F1	C/D	(67, 68)
YY1	C/D	(87)
MDM2	C/D	(88)
c-Jun	C/D	(72, 73, 89–91)
AP-2	C/D	(92)
myc	C/D	(74)
NFAT-1	C/D	(71)
Huntington disease (HD) protein	C/D	(93, 94)
Cyclin A	C/D	(95)
Oct-1	A/B	(96)
TATA binding protein (TBP)	A/B	(25)
Hepatocyte nuclear factor 3 (HNF3)	nd[a]	(97)
HNF-4	D	(98)
p53	D	(65, 99–101)
Myocyte enhancer factor 2C (MEF2C)	A/C	(102)
SMAD2, SMAD3, and SMAD4	C/D	(75, 76, 103–107)
Msx1	nd	(108)
Viral proteins (c-rel, p-52, p-50, rel A, tat, BPV-E2)	nd	(109–111)
E1a	C/D	(112, 113)
Retinoblastoma (Rb) protein	nd	(114, 115)
p107 (Rb-like protein)	nd	(116)
DNA methyltransferase 1 (DNMT1)	nd	(117)
Zinc finger repressor protein 89 (ZBP89)	nd	(118)

[a]nd, not determined.

been indicated, Sp1, Sp3, and possibly other Sp/KLF proteins are important for basal activity of p21 through interactions with six proximal GC-rich motifs (Fig. 3). Several reports also show multiple cell context-dependent pathways for activation of p21 through direct interactions of other transcription factors with Sp1. These include p53 (DNA-dependent) activation primarily through GC-rich site 3 in several cell lines (65, 66); calcineurin-induced NFAT-1 and NFAT-2 (Sp1) activation of GC-rich sites 1–6 (71); c-jun activation through GC-rich sites 1–6 in HepG2 cells (72); repression of p21 expression in human embryonic epithelial 293 cells through GC-rich site 3 (73); c-myc-induced repression in Caco-2 colon adenocarcinoma cells through GC-rich sites 1–6 (74); and transforming growth factor β (TGFβ) effectors SMAD2, SMAD3, and SMAD4 activation through GC-rich sites 1–6 (75, 76). These data, coupled with the reported direct p53-GKLF-4 interactions at GC-rich site 1 (66) and indirect coactivation by p300 (30), illustrate the highly flexible role of Sp1 and other Sp/KLF proteins in modulating expression of a single gene.

Table I summarizes the growing list of transcription factors that bind Sp1 or Sp/KLF proteins and modulate (enhance or decrease) gene expression (25, 65–67, 67, 68, 68–118), including the AhR, Arnt, GATA-1, GATA-2, GATA-3, NF-YA, VHL, MyOD, HDAC1, PML, HTLF, E2F1, YY1, MDM2, c-jun, AP-2, myc, NFAT-1, HD protein, cyclin A, Oct-1, TBP, HNF3, HNF4, p53, MEF2C, SMAD2, SMAD3, SMAD4, Msx1, several viral proteins (c-rel, p50, p52, rel A, tat, and BPV-E2), El1, Rb, p107 (Rb-like), DNMT1, and ZBP-89. The domain-specific interactions of Sp1 and other transcription factors have been investigated in several studies and the C-terminal (C/D) domain is the major site for protein interactions with Sp1 (Fig. 1). Surprisingly, protein interactions with other domains of Sp1 have been detected for only a small number of transcription factors, suggesting that competitive binding of proteins at the C-terminal domain of Sp1 may be an important regulatory pathway. This is currently being investigated in this laboratory.

IV. Interactions of Sp Family Proteins with Estrogen Receptors

A. DNA-Dependent Estrogen Receptor-Sp1 Interactions

The estrogen receptor (ER) is a ligand-induced transcription factor and a member of the nuclear receptor (NR) superfamily. Other members of this family include the steroid hormone receptors, thyroid hormone, vitamin D and peroxisome proliferator-activated receptors (PPARs), retinoid acid receptor (RAR), retinoid X receptor (RXR), and several orphan receptors (119–123). NRs, such as the ER, contain several common structural domains (Fig. 4) including an N-terminal A/B domain, which contains activation function 1

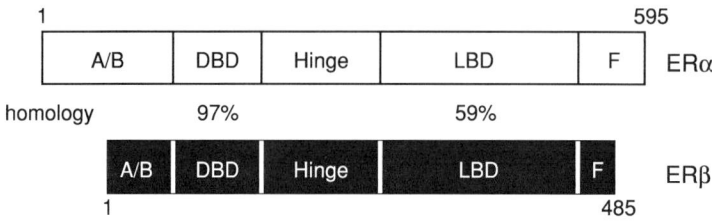

FIG. 4. Structures of ERα and ERβ. Both proteins contain activation function 1 (AF-1) (A/B), DNA binding (C), hinge (D), AF-2, and ligand binding (E/F) domains.

(AF1), a DNA-binding domain (C) containing two zinc fingers (ZF1 and ZF2), a hinge domain (D), the ligand binding domain E, which also contains AF2 and an N-terminal F region. ER-mediated transactivation is classically represented as a ligand-induced formation of an ER homodimer which interacts with consensus palindromic estrogen responsive elements (EREs) (GGTCA(N)$_3$-TGACC). Many E2-responsive gene promoters also contain functional nonconsensus EREs. ERα was the first ER identified; however, ERβ has also been characterized and both forms are differentially expressed in multiple tissues (124–127). Therefore, it is possible that ERα:ERβ heterodimers may also be ligand-induced transcription factors (128).

Several studies have identified E2-responsive gene promoters that do not contain EREs but contain ERE half-sites and GC-rich sites (129, 130). For example, the E2-responsive regions of the c-myc and creatine kinase B (CKB) gene promoters contain GGGCA(N)$_{16}$GGCGG and GGTCA(N)$_{21}$GGCGG motifs, respectively, and constructs containing c-myc and CKB promoter inserts were E2-responsive in HeLa cells transfected with ERα. In contrast, cotransfection with a DNA binding domain deletion mutant of ERα (HE11) did not result in hormone-induced transactivation, suggesting that DNA binding of ERα was required. Rochefort and coworkers also characterized the E2-responsive region (131, 132) of the cathepsin D gene promoter, which contained multiple nonconsensus ERE half-sites and GC-rich motifs. Deletion analysis studies with constructs containing cathepsin D gene promoter inserts identified an E2-responsive region (-199 to -165) containing a GGGCGC(N)$_{23}$ACGGG motif (133, 134). This promoter was extensively analyzed by mutation analysis in transient transfection studies and in gel mobility shift assays, and results indicated that hormone-induced transactivation required the GC-rich and nonconsensus ERE half-site.

Results illustrated in Fig. 5A show that constructs containing the GC(N)$_x$ERE1/2 motifs from the cathepsin D and TGFα gene promoters were E2-responsive, and mutation of the GC-rich sites resulted in loss of E2-responsiveness (133–135). Initial studies with Hsp27 used a construct with

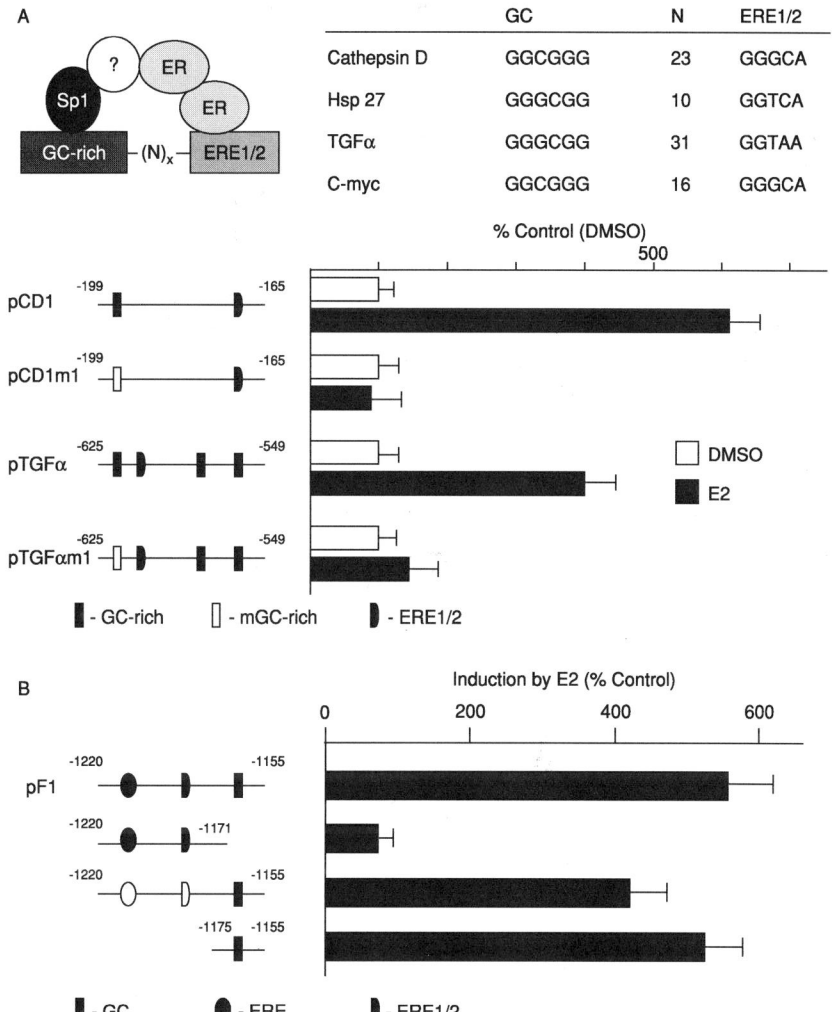

FIG. 5. ERα/Sp1-mediated transactivation. (A) DNA-dependent ERα/Sp1. The E2-responsive GC(N)$_x$ERE1/2 motifs in the cathepsin D, Hsp 27, TGFα, and c-myc gene promoters are indicated. Hormone-dependent activation of wild-type and mutant pCD1 and pTGFα constructs derived from the cathepsin D and TGFα gene promoters are dependent on GC-rich motifs (133–135).

mutations in both GC-rich and ERE1/2 sites, and the mutant construct was not E2-responsive in transactivation studies (136). Interactions of nuclear extracts from breast cancer cell lines with oligonucleotides containing both the GC-rich and ERE half-sites from the cathepsin D, TGFα, and Hsp27 gene promoters gave a broad retarded band in gel mobility shift assays. The intensity of this band was decreased by competition with excess unlabeled consensus GC-rich or ERE oligonucleotides, and Sp1 or ERα antibodies typically supershifted or decreased (immunodepletion) the intensity of the retarded band. Although these data were consistent with formation of an ER/Sp1 complex where both transcription factors bound promoter DNA, incubation of oligonucleotides with recombinant ERα and Sp1 proteins did not give the typical broad retarded band complex. These data suggest that other factors expressed in nuclear extracts from breast cancer cells are required for formation of the ERα/Sp1 complex. In contrast, the E2-responsive GC(N)$_x$ERE1/2 motif in the progesterone receptor gene (+565 to +601) bound ERα and Sp1 proteins (additively), but did not form a cooperative ER/Sp1-DNA complex (137).

The hormone-responsiveness of the rabbit uteroglobin gene also involves interactions of ERα and Sp1 proteins (138–140). Genomic footprinting of the proximal −300 to −200 promoter region of this gene showed that after treatment with E2, there was occupation of a nonconsensus ERE and an adjacent GC-box (−270 to −21), whereas downstream GC-boxes were not affected. Results of transient transfection studies with constructs containing mutant ERE and GC-rich sites also demonstrated that both the ERE and adjacent GC-rich site were required for hormone-dependent transactivation. Incubation of nuclear extracts from endometrial endothelial cells with an oligonucleotide containing the ERE and GC-rich site did not form an ERα/Sp1 complex in gel mobility shift assays, but showed retarded bands corresponding to ERα-Sp1-DNA complexes. In contrast, titration experiments with ERα and Sp1 proteins gave a supershifted higher molecular weight retarded band in gel mobility shift assays. These results suggest that the ERE(N)$_6$GC/T motif (*GGTCAC* CA *TGCCC* TGGCTT *GCCACACCC*) within the rabbit uteroglobin gene promoter represents another E2-responsive element where both ERα and Sp1 cooperatively interact. The *Xenopus* vitellogenin A1 gene contains a downstream (i) and upstream GC-rich (io) promoter separated by 1.8 kb (141). The E2-responsive i promoter contains multiple EREs, whereas E2-dependent transactivation of the io promoter requires both the GC-rich motifs as well as the downstream EREs. These data suggest that ERα/Sp1-mediated transactivation from the io promoter is dependent on GC and ERE sites that are widely separated. The nature of this complex interaction requires further investigation.

B. ERα/Sp1 Activation of GC-Rich Promoters

The GC(N)$_x$ERE1/2 motifs identified in the TGFα, cathepsin D, and Hsp27 gene promoters are also present in several E2-inducible genes such as c-*fos* protooncogene that are associated with hormone-dependent growth of

ER-positive breast cancer cells (142). Previous studies had identified a nonconsensus ERE (*CGGCA GCG TGACC*) in the distal region (-1212 to -1200) of the c-*fos* gene promoter that bound ER in gel mobility shift assays. A single copy of this region of the promoter was not hormone-responsive in transient transfection assays in HeLa cells, whereas a construct containing three tandem copies was activated by E2. Closer examination of this region of the c-*fos* promoter identified an additional ERE1/2 and a nonconsensus 5'-GGGCGTGG-3' Sp1 binding site at -1168 to -1161. Transient transfection studies with a c-*fos* promoter construct containing the Sp1 binding site were E2-responsive (Fig. 5B), suggesting that ERα/Sp1 interactions with one or more $GC(N)_x ERE1/2$ or $GC(N)_x ERE$ motifs may be required for hormone-induced transactivation (143). However, results of mutation analysis demonstrate that the nonconsensus ERE and ERE1/2 sites are not required for activation by E2, whereas the GC/GT site alone was sufficient for hormone-responsiveness (Fig. 5B). Moreover, reexamination of the Hsp27 gene promoter also showed that constructs containing only the GC-rich motif were induced by E2 in transient transfection assays, and similar results were observed with constructs containing 1 or 3 consensus GC-rich promoters (144).

Further analysis of ERα and Sp1 protein binding to GC-rich oligonucleotides showed that Sp1, but not ERα, directly bound these elements, and ERα/Sp1-DNA supershifted complexes were not detected in gel mobility shift assays. Kinetic analysis studies showed that ERα enhanced the on-rate of Sp1-DNA binding, but did not affect dissociation (off-rate) of the Sp1-DNA complex (144). Similar results have been observed for other interacting transcription factors. For example, sterol regulatory element-binding protein, cyclin D1, and the human T cell lymphotropic virus type 1 Tax protein enhance DNA binding of Sp1, ER, and other DNA-bound nuclear proteins, respectively, without forming supershifted complexes (145–151). Subsequent studies showed that ERα and Sp1 can be coimmunoprecipitated, and pulldown studies using chimeric glutathione-S-transferase-Sp1 (wild-type/variant) proteins demonstrated that ERα interacts with the C-terminal DNA-binding domain of Sp1 (144, 152). The interaction of ERα and many other transcription factors with a common region of Sp1 (Table I) suggests that this binding site may exhibit functional significance for proteins which compete for binding to Sp1.

GC-rich sites that bind Sp1 and other Sp/KLF proteins are present in most mammalian gene promoters, including many E2-responsive genes involved in cell cycle progression, nucleotide biosynthesis, and metabolism. Research in this laboratory has investigated the potential role of ERα/Sp1 interactions with specific GC-rich motifs in mediating E2-induced transactivation. Most of these studies were carried out in ER-positive breast cancer cells and, in some cases, in ER-negative cell lines. GC-rich promoters in pCDNA3 are weakly

E2-responsive in ER-positive cells, and induced activity is observed only after cotransfection with ERα expression plasmid. The requirement for ERα cotransfection in breast cancer cells has been observed in studies with constructs containing many other E2-responsive gene promoters. This is due, in part, to limiting levels of ERα in transfected cells that overexpress the promoter constructs (*129, 130, 142, 153, 154*). Figure 6 summarizes results of several studies of E2-responsive genes which have shown that one or more GC-rich elements are required for activation by E2 (*56–58, 143, 155–163*).

Studies in this laboratory have focused on genes induced by E2 in ER-positive MCF-7, T47D, or ZR-75 breast cancer cells. Results obtained for low density lipoprotein receptor (LDLR), epidermal growth factor receptor (EGFR), rat SK3 (rSK3–a small conductance Ca^{+2}-activated potassium channel), and receptor for advanced glycation end products (RAGE) were primarily determined in non-breast cancer cell lines (*164–167*). A recent study showed that hormone-responsiveness of the progesterone receptor gene (promoter B) was also dependent on ERα/Sp1 interactions with GC-rich site (−80/−34 region) (*168*). The hormone-responsive GC-rich sites are primarily located in proximal promoter regions; however, more distal GC-rich motifs

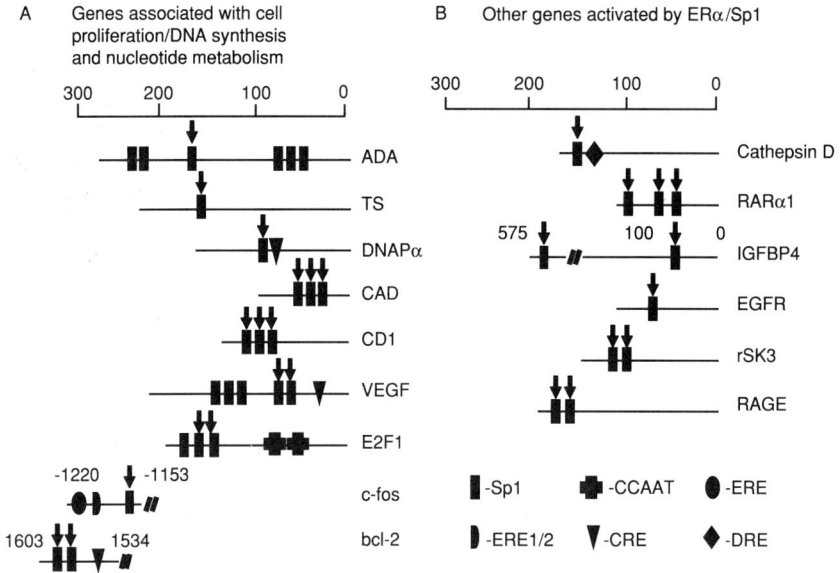

FIG. 6. Summary of ERα/Sp1-regulated genes. E2-responsive GC-rich motifs in multiple gene promoters activated by ERα/Sp1 are indicated by an arrow (*56–58, 143, 155–167*).

have been identified in the c-fos, bcl-2, and insulin-like growth factor binding protein 4 (IGFBP-4) promoters (Fig. 6). Hormonal activation of constructs containing adenosine deaminase promoter inserts required only one of the six proximal GC-rich sites, whereas multiple sites are active in several other promoters. The precise contributions of individual GC-rich sites within various promoters to E2-responsiveness has not been unequivocally defined. Promoter analysis has demonstrated that many genes activated through genomic ERα/Sp1 are also coordinately upregulated via nongenomic pathways of E2 action. For example, cyclin D1, c-*fos*, *bcl*-2, and E2F1 genes are induced by ERα/Sp1 and hormone activation of cAMP, phosphatidylinositol-3-kinase (PI3K), and mitogen-activated protein kinase (MAPK) pathways (*58, 159, 162, 169*). Cathepsin D not only contains E2-responsive $GC(N)_{23}ERE$ and GC-rich motifs, but also a downstream nonconsensus ERE activated by ERα and USF1/2 (*133, 134, 155, 170*). A recent study in this laboratory also demonstrates different mechanisms of hormonal activation of E2F1 gene expression in MCF-7 and ZR-75 breast cancer cells (*58*). In the former cell line, both GC-rich and NFY sites are required for E2-induced transactivation through a cooperative ERα/Sp1-NFY complex. In contrast, the GC-rich and NFY sites in ZR-75 cells are independently activated by E2 through ERα/Sp1 (genomic) and cAMP-dependent phosphorylation of NFYA (nongenomic), respectively.

The results summarized in Fig. 6 show that many genes required for E2-dependent breast cancer cell growth are regulated via ERα/Sp1. The importance of Sp1 in E2-induced proliferation of MCF-7 cells was investigated by RNA interference using small inhibitory RNA for Sp1 (iSp1) and determining the effects on cell cycle progression. iSp1 inhibits basal and E2-induced $G_0/G_1 \rightarrow S$ phase cell cycle progression, further demonstrating an important role for ERα/Sp1-dependent genes in the growth of ER-positive breast cancer cells (*171*).

Many E2-responsive gene promoters, such as cathepsin D, TGFα, c-*fos*, creatine kinase B, and RARα, contain both GC-rich and ERE/ERE1/2 motifs; however, their mechanisms of hormone-induced transactivation can only be ascertained by rigorous promoter analysis. The c-fos, CKB, and RARα gene promoters contain $GC(N)_xERE/ERE1/2$ motifs, but in breast cancer cells, hormone-induced transactivation is due to interactions of ERα/Sp1 with GC-rich elements. In contrast, both ERα and Sp1 bind $GC(N)_xERE1/2$ motifs in the cathepsin D and TGFα promoters (*133–135, 155*), whereas in the prothymosin gene (*172*), which also contains GC-rich/ERE1/2 sites, hormone-dependent activation of ERα is Sp1-independent. These data clearly demonstrate the importance of promoter context in determining pathways for estrogen action. However, other factors that are important for cell context-dependent selective ERα/Sp1 interactions with $GC(N)_xERE1/2$ or GC-rich motifs have not yet been defined.

C. Comparative Activation of GC-Rich Promoters by ERα/Sp1 and ERβ/Sp1

ERα/Sp1 activation of genes through interactions with GC-rich motifs represents an important pathway for hormone-induced transactivation in breast cancer cells (Fig. 6). ERα/Sp1 action does not require binding of ERα to promoter DNA, and this response is also observed in cells transfected with DNA binding domain deletion mutants of ERα (HE11). ERα/AP1 is another example of a hormone-induced pathway where ERα does not bind DNA but interacts with c-jun, which is part of the AP1 complex (*173–177*). ERα/AP1 and ERβ/AP1 are differentially activated by estrogens and antiestrogens in various cell lines, and we also investigated the comparative activation of ERα/Sp1 vs ERβ/Sp1 (*152*). Results in Fig. 7 show that E2 activates ERα/Sp1 in breast and prostate cancer cells but not in HeLa cells, whereas hormone-dependent activation of ERα/Sp1 was not observed in any of the cell lines and decreased activity was observed in HeLa cells. ERβ coimmunoprecipitates with Sp1 and like ERα interacts with the C-terminal domain of Sp1, suggesting that differences in ERα/Sp1 and ERβ/Sp1 action may be related to the structures of the ER subtypes. Saville and coworkers (*152*) constructed chimeric proteins in which the A/B (AF1) domains of ERα and ERβ were fused to the C–F domains of ERβ and ERα, respectively. The results (Fig. 7) show that the A/B domains were critical determinants of ER/Sp1-mediated transactivation, and the A/B domain of ERβ lacked the required transactivation function. Other studies on the A/B domains of ERβ and ERα suggest that AF1 is significantly weaker in ERβ, and this is probably due to less favorable interactions with critical coregulatory proteins.

D. Domains of ERα Required for Estrogen and Antiestrogen Activation of ERα/Sp1

E2 and the antiestrogens 4′-hydroxytamoxifen (4-OHT) and ICI 182,780 activate reporter gene activity in cells transfected with constructs containing GC-rich promoters (pSp1 or pSp3) (*152*). Initial studies focused on regions within the A/B domain of ERα required for ERα/Sp1 action, and it was shown that amino acids 79–177 were sufficient for ligand-induced transactivation. In studies on the low-density lipoprotein gene promoter, amino acids 67–139 were critical for ERα/Sp1 activation (*166*); however, the specific amino acids required for this response were not identified in either study. The contributions of the C–F domains to estrogen and antiestrogen activation of ERα/Sp1 have revealed a complex ligand-dependent pattern (Fig. 8) (*178*). ERα/Sp1 activation of GC-rich promoters was independent of the DNA binding domain of ERα, and the results illustrated in Fig. 8A compare ligand-induced transactivation in cells expressing ERα variants containing deletions of zinc finger 1 (aa 185–205)

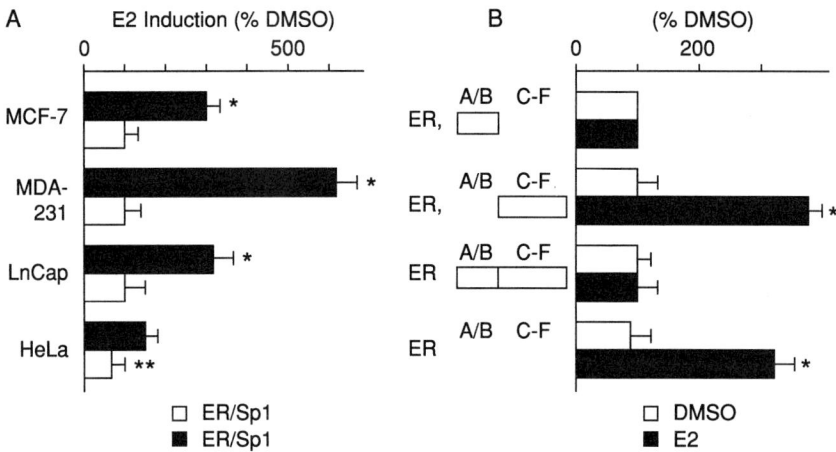

FIG. 7. ER/Sp1 activation of pSp1$_3$ in multiple cell lines. (A) Activation by ERα/Sp1 and ERβ/Sp1. Different cell lines were transfected with pSp1$_3$ and ERα or ERβ expression plasmids, treated with DMSO or 10 nM E2. The fold induction of luciferase activity in each cell line is given and significant (p < 0.05) induction (∘) or inhibition (∘∘) by E2 is indicated. (B) Activation of pSp1$_3$ by chimeric ER in MDA-MB-231 cells. ERα, ERβ, and chimeric receptors where AF-1 domains are interchanged were used in this study. Significant (p < 0.05) induction by E2 was observed only in cells transfected with ERα or chimeric ERα/β (152).

(ERαΔZF1) or zinc finger 2 (aa 218–245) (ERαΔZF2). The results show that E2, but not antiestrogen 4-OHT or ICI 182780, activate ERαΔZF1/Sp1 or ERαΔZF2/Sp1; however, in cotreatment experiments, the antiestrogens inhibited E2-induced transactivation. These results suggest that ERα/Sp1 activation by antiestrogens requires the DBD of ERα, and this may involve direct or indirect interactions with regulatory proteins.

Regions within the D–F domains of ERα required for estrogen and antiestrogen activation of ERα/Sp1 were also investigated and shown to be highly dependent on ligand structure (Fig. 8B). Deletion of aa 271–300 in the hinge region did not affect ERα/Sp1 activation by estrogens or antiestrogens, whereas a mutant containing a deletion of helix 1 and part of helix 2 (hERαΔ265–330) was not activated by estrogens or antiestrogens. D538N, E542Q, and D45N mutations in helix 12 of the E domain of ERα (ERαTAF1) do not affect activation of ERα/Sp1 by estrogens or antiestrogens; in contrast, these amino acids within helix 12 are important for E2-dependent activation of ERE promoters (pERE$_3$) by ERα and recruitment of p160 steroid receptor activators (SRCs) (119–123).

Further deletion analysis of the E/F domain of ERα showed that there were dramatic differences in the domain requirements for activation of

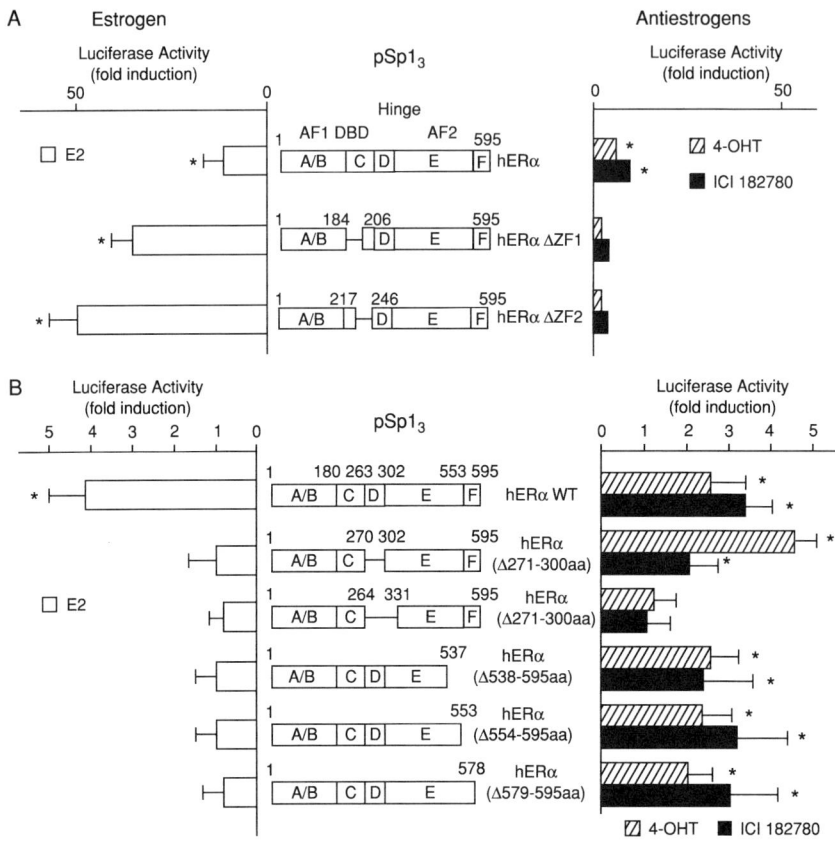

FIG. 8. ERα domain requirements for estrogen and antiestrogen activation of ERα/Sp1 in MCF-7 cells. MCF-7 cells were treated with estrogens or antiestrogens, transfected with pSp1₃, wild-type, or mutant ERα expression plasmids, and luciferase activity was determined. Significant ($p < 0.05$) induction is indicated (∘) (178).

ERα/Sp1 by estrogens and antiestrogens. Deletion of amino acids 537–595, 554–595, or 578–595 did not affect antiestrogen activation of ERα/Sp1, whereas E2 was inactive using all of these deletion mutants. These data suggest that the C-terminal region of the F domain of ERα (aa 578–595) is required for activation of ERα/Sp1 by E2, whereas deletion of the F domain does not affect ERα-dependent transactivation of pERE₃. This was confirmed in experiments using GAL4-peptide chimeras in which the peptides contain NR boxes (LXXLL sequences) or the Fβ peptide (aa 575–595) of ERα (178). Expression of the Fβ peptide inhibited E2-induced transactivation in cells transfected with pSp1₃ but not pERE₃, whereas expression of the NR box

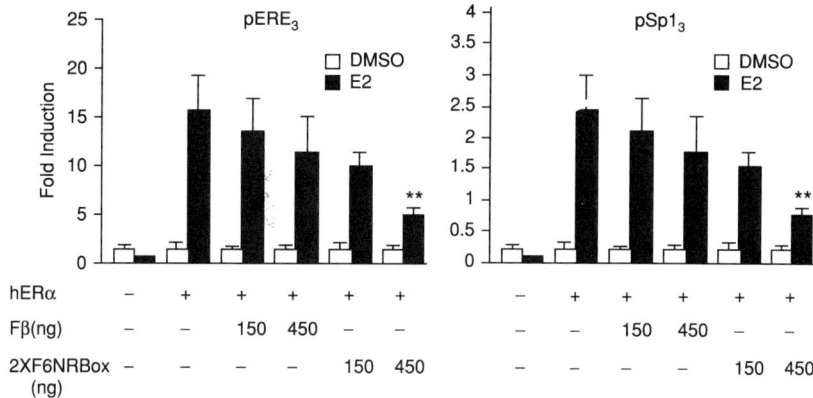

FIG. 9. Fβ peptide decreases ERα/Sp1-mediated transaction. Cells were treated with E2 or DMSO, transfected with pERE or pSp1$_3$ and GAL-fusion proteins with an NR-box peptide or F domain peptide (178). Significantly ($p < 0.05$) decreased E2-induced luciferase activity is indicated (∘∘).

peptide inhibited transactivation in cells transfected with pERE$_3$ but not pSp1$_3$ (Fig. 9). Thus, estrogen or antiestrogen activation of ERα/Sp1 requires different surfaces of ERα, which subsequently interact with other nuclear coregulatory proteins.

E. ERα/Sp3 Interactions with GC-Rich Promoters

Vascular endothelial growth factor (VEGF) plays an important role in angiogenesis and is induced by multiple factors including steroid hormones. For example, VEGF is induced by E2 in the rodent uterus and in some breast/endometrial cancer cells (179–182). Initial studies in this laboratory showed that E2-dependent induction of VEGF mRNA levels and reporter gene (luciferase) activity in cells transfected with VEGF promoter constructs was highly variable in breast and endometrial cancer cell lines. E2 decreased VEGF mRNA levels in HEC1A endometrial cancer cells and luciferase activity in cells transfected with several VEGF promoter constructs (183). Hormone-dependent decreases observed in transient transfection studies were mapped to the proximal GC-rich sites (-131 to -52) which primarily bound Sp1 and Sp3 in gel mobility shift assays. Dominant negative Sp3 reversed E2-dependent downregulation of VEGF promoter constructs, suggesting that ERα/Sp3 was required for this response. In contrast, E2 induced VEGF mRNA levels in ZR-75 breast cancer cells and E2-responsiveness was associated with the same proximal GC-rich region of the VEGF promoter (184). A distal ERE in the VEGF promoter was linked to E2-responsiveness in Ishikawa endometrial cancer cells (*182*); however, this element was not required in ZR-75 cells. RNA interference studies with small inhibitory RNAs for Sp1 and Sp3 showed that E2-induced transactivation of a VEGF promoter construct containing the GC-rich sequence was decreased by silencing Sp1 or Sp3. These results suggested that E2-induced transactivation of VEGF in ZR-75 cells requires both ERα/Sp1 and ERα/Sp3, indicating that ERα/Sp3 mediates cell context-dependent up- or downregulation of VEGF by E2.

F. Coactivation of ERα/Sp-Dependent Transactivation

ERα-mediated transactivation from ERE promoters requires ligand-dependent homodimer formation and interaction with appropriate *cis*-elements. This process is accompanied by recruitment of nuclear coactivators and coregulators that enhance transactivation and interactions with the basal transcription machinery (*185–192*). The p160 steroid receptor coactivators (SRCs) exhibit ligand-dependent interactions with the AF2 (E) domain of ERα and other NRs, and these coactivators may also exhibit histone acetyltransferase activity. Subsequent studies suggest that there are multiple classes of coregulatory proteins, and enhanced ERα-dependent transactivation may involve coordinate coactivation by different classes of these proteins (*192*). Prototypical SRCs that interact with ERα and other NRs through their LXXLL motifs (NR boxes) do not coactivate ERα/Sp1 in breast cancer cells. Their predominant effect on ERα/Sp1 is inhibitory and resembles the action of corepressors (*178*). Ongoing studies in this laboratory have identified several mediator complex proteins, such as vitamin D interacting protein 205 (DRIP 205), DRIP 150 and DRIP 130, and the RING protein SNURF, as coactivators of ERα/Sp1; however, their mechanisms of coactivation are currently unknown and are being investigated.

V. Interaction of Sp Family Proteins with Other Steroid Hormone Receptors and NRs

A. Progesterone Receptor (PR)/Sp1-Mediated Transactivation

Treatment of T47D cells with progesterone induces cell proliferation followed by growth arrest. In transient transfection assays, progesterone also induces a p21 promoter construct (*193*). Subsequent deletion analysis shows that progestin-responsiveness is associated with GC-rich sites 3 and 4 in the p21 promoter (Fig. 3). PR coimmunoprecipitates with Sp1 and p300, and it was proposed that p300 coregulates PR/Sp1 activation of the p21 promoter through proximal GC-rich motifs. There is also evidence that in HEC-1 endometrial cancer cells PR-A/Sp1 and PR-B/Sp1 activate the glycodolin promoter through GC-rich sites. Basic transcription element binding protein

(BTEB) is an Sp/KLF family member and preliminary studies suggest that PR/BTEB may be required for activation of the uteroferrin gene in endometrical cancer cells. Specific elements within this gene promoter were not defined (*194*).

B. Androgen Receptor (AR)/Sp1-Mediated Transactivation

Androgens also induce p21 expression which is due, in part, to an androgen-responsive element at −215 in the p21 promoter (*195*). However, reexamination of the p21 promoter in LNCaP cells showed that in transient transfection assays, downstream proximal GC-rich sequences were also androgen-responsive. Deletion analysis indicated that GC-rich site 3 was primarily responsible for androgen-dependent transactivation, and limited studies indicated that the ligand binding domain of AR was sufficient for activation of a GC-rich p21 promoter construct. AR and Sp1 are also coimmunoprecipitated in LNCaP cells; however, the interacting domains were not determined. Androgens suppress the upregulation of rat luteinizing hormone β (LHβ) by GnRH, and deletion analysis of the LBβ promoter indicated that decreased transactivation was associated with GC-rich and possibly egr-1 sites (*196*). This study also confirmed direct AR/Sp1 interactions and the requirement of the DNA binding domain of AR for AR/Sp1 complex formation. In contrast, the ligand binding E/F domain of AR was sufficient for AR/Sp1-dependent activation of the GC-rich p21 promoter, suggesting that functional interactions of these transcription factors are complex and dependent on cell and promoter context.

C. Retinoid Receptor/Sp1-Mediated Transactivation

A GC-rich Sp1 binding site in the interleukin 1B (IL-1B) gene promoter was sufficient for upregulation by retinoic acid (*197*). They reported that several nuclear receptors including the RAR, RXR, thyroid hormone receptor (TR), vitamin D receptor (VDR), and PPAR$_\gamma$ interact with Sp1 to give a higher molecular weight ternary complex in gel mobility shift assays. This complex was not previously observed for ER/Sp1 but this could be a function of assay conditions. In addition, Sp1 enhanced RARα binding to a retinoic acid response element in gel mobility shift assays. Both RARα and RXR interact with the C-terminal domain of Sp1, as previously observed with many other nuclear transcription factors (Table I). Suzuki and coworkers (*198*) also reported that urokinase plasminogen activator (uPA) was induced by retinoid acid in bovine aortic endothelial cells, and deletion analysis of the uPA promoter identified a retinoid-responsive GC-rich site. Their results also showed that RXR and RAR coimmunoprecipitated with Sp1 and enhanced Sp1-DNA binding in gel mobility shift assays. A supershifted ternary complex was not detected. Subsequent studies (*199*) in the same cell line also showed that other retinoid-responsive genes including transglutaminase, TGFβ1, and TGFβ receptors I and II were activated through RAR:RXR/Sp1 interactions with GC-rich motifs. Analysis of individual GC-rich promoters demonstrated sequences flanking the core GC-rich motifs markedly influenced both Sp1-DNA binding and transactivation. For example, GC boxes that contain 5 tandem purine or pyrimidine bases at the 3′-flanking region of GC-rich motifs exhibited minimal Sp1 binding. These results may be important for Sp1- and NR/Sp1-mediated transactivation of multiple genes/gene promoters and require further characterization and study in multiple cell lines.

D. PPARγ/Sp1-Mediated Transactivation

The classical mechanisms of PPARγ-mediated transactivation involves interaction of the PPARγ:RXR heterodimer with PPARγ response elements in target gene promoters. However, repression of thromboxane receptor gene expression by PPARγ agonists was shown to be dependent on a −22 to −7 GC box, and treatment of vascular smooth muscle cells decreased formation of an Sp1-DNA complex using nuclear extracts from this cell line (200). PPARγ and Sp1 interacted in a pulldown assays; however, the interacting domains were not determined. Thus, PPARγ inhibitory action may be related to sequestering of Sp1, thereby decreasing Sp1-mediated transactivation.

E. Chicken Ovalbumin Upstream Promoter-Transcription Factor (COUP-TF)/Sp1-Mediated Transactivation

COUP-TF is an orphan nuclear receptor that modulates expression of multiple genes, and induction of human immunodeficiency virus type 1 (HIV-1) long terminal repeat and Egr-1/NFG1-A has been linked to COUP-TF and GC-rich promoter elements (201, 202). For example, in microglial cells, COUP-TF-dependent regulation of HIV-1 was associated with a GC-rich promoter sequence (−68 to +29) that bound nuclear extracts to form Sp1/Sp3-DNA complexes but not a supershifted ternary band. However, COUP-TF and Sp1 physically interact through the DNA binding domain of COUP-TF. Pipaón and coworkers (202) also showed similar COUP-TF-Sp1 interactions which were important for COUP-TF/Sp1-mediated transactivation of Egr-1 in a rat urogenital mesenchymal cell line through interactions with the GC-rich −64 to −46 region of the Egr-1 promoter.

F. Steroidogenic Factor-1 (SF-1)/Sp1-Mediated Transactivation

SF-1 is an orphan nuclear receptor that is preferentially expressed in the anterior pituitary gland, adrenal gland, and gonads, and regulates tissue-specific gene expression through SF-1 binding sites. SF-1 is important for cAMP-dependent expression of the bovine cholesterol side-chain cleavage enzyme (P450scc or CYP11A) in adrenal cells, and both SF-1 and Sp1 cooperatively upregulate CYP11A in bovine luteal cells and Y-1 cells (203). SF-1 and Sp1 interact and Sp1 specifically binds the N-terminal and DNA-binding domain of SF-1. Gel mobility shift data suggest that Sp1 and SF-1 bind DNA to form a ternary complex. Sp1 and SF-1 also cooperatively play a role in cAMP-dependent activation of human steroidogenic acute regulatory (StAR) protein gene expression in Y-1 mouse adrenal tumor cells (204). Results of promoter analysis and gel mobility shift assays clearly demonstrate Sp1-SF-1 interactions; moreover, using an SF-1 oligonucleotide, a ternary complex was observed. This study suggests that Sp1-SF-1 interactions with DNA are primarily dependent on the SF-1 motif, suggesting that Sp1/SF-1-dependent activation of StAR may involve Sp1/SF-1-DNA binding in cooperation with other nuclear coregulatory proteins.

VI. Summary

The Sp/KLF family of nuclear transcription factors plays an important role in regulating gene expression in multiple cell lines. Constitutive expression of several genes is dependent on Sp1 and other family members which interact with proteins associated with the basal transcription machinery, including TAFs, TBP, and mediator complex proteins. Another level of transcriptional control involves cell context-dependent differences in expression of Sp/KLF proteins and their competitive binding to specific GC-rich promoters. These interactions can result in additive, synergistic, or antagonistic expression of genes and may depend, in part, on relative cellular expression of Sp proteins and their interactions with *cis* promoter elements. For example, Sp1 and Sp3 can cooperatively activate gene expression or Sp3 can inhibit Sp1-dependent transactivation, and these interactions are promoter- and cell context-dependent. Interactions of Sp1 with other DNA-bound transcription factors including ERα have been extensively described and are also important for gene regulation.

The major focus of this review has been hormone-dependent activation of ERα/Sp1 through specific GC-rich promoter sequences and the contributions of this pathway to growth of ER-positive breast cancer cells. The results clearly demonstrate that this nonclassical genomic pathway may be a critical determinant for hormone-dependent cell growth. It was also evident that the mechanism of ERα/Sp1-mediated transactivation was different from that described for activation of ERα on ERE promoters. Domains of ERα required for ERα/Sp1 and ERα (on an ERE)-mediated transactivation and the effect of ER ligands (e.g., estrogens and antiestrogens) are different. Moreover, the p160 coactivators that exhibit ligand-dependent interactions with the AF2 region of ERα and coactivate ERα (on an ERE) inhibit ERα/Sp1-mediated transactivation in breast cancer cells (178). These results suggest that the as yet unidentified coactivators of ERα/Sp1 may be critical factors for growth of breast cancer cells, and the identities of coactivators and mechanisms of ERα/Sp1 coactivation are currently being investigated. It is also apparent that other nuclear receptors activate gene expression through interactions with Sp1.

Differences in cell phenotypes are related, in part, to differences in their patterns of gene expression which, in turn, are dependent on coordinated expression of transcription factors and other nuclear coregulatory proteins. Sp1/KLF proteins typify the increasing complexities that have been revealed for this and other families of transcription factors. Sp proteins not only regulate basal expression of diverse mammalian and viral genes, but also form interacting networks with many other nuclear proteins to modulate gene expression. The studies with ERα/Sp1 have revealed a unique genomic mechanism of ERα action in breast cancer cells, and reports from several laboratories show that other members of the NR family also act through NR/Sp protein complexes. Current research is focused on contributions of this pathway in mediating gene expression in normal cells/tissues using both *in vitro* and *in vivo* models.

Acknowledgments

This work was supported by the National Institutes of Health (ES09106 and CA104116), the Texas Agricultural Experiment Station, and the Sid Kyle Endowment.

References

1. Dynan, W. S., and Tjian, R. (1983). The promoter-specific transcription factor Sp1 binds to upstream sequences in the SV40 early promoter. *Cell* **35,** 79–87.
2. Dynan, W. S., and Tjian, R. (1985). Control of eukaryotic messenger RNA synthesis by sequence-specific DNA-binding proteins. *Nature* **316,** 774–778.
3. Dynan, W. S., and Tjian, R. (1983). Isolation of transcription factors that discriminate between different promoters recognized by RNA polymerase II. *Cell* **32,** 669–680.
4. Gidoni, D., Dynan, W. S., and Tjian, R. (1984). Multiple specific contacts between a mammalian transcription factor and its cognate promoters. *Nature* **312,** 409–413.
5. Gidoni, D., Kadonaga, J. T., Barrera-Saldana, H., Takahashi, K., Chambon, P., and Tjian, R. (1985). Bidirectional SV40 transcription mediated by tandem Sp1 binding interactions. *Science* **230,** 511–517.
6. Briggs, M. R., Kadonaga, J. T., Bell, S. P., and Tjian, R. (1986). Purification and biochemical characterization of the promoter-specific transcription factor, Sp1. *Science* **234,** 47–52.
7. Kadonaga, J. T., Carner, K. R., Masiarz, F. R., and Tjian, R. (1987). Isolation of cDNA encoding transcription factor Sp1 and functional analysis of the DNA binding domain. *Cell* **51,** 1079–1090.
8. Kadonaga, J. T., Courey, A. J., Ladika, J., and Tjian, R. (1988). Distinct regions of Sp1 modulate DNA binding and transcriptional activation. *Science* **242,** 1566–1570.
9. Courey, A. J., and Tjian, R. (1988). Analysis of Sp1 *in vivo* reveals multiple transcriptional domains, including a novel glutamine-rich activation motif. *Cell* **55,** 887–898.

10. Courey, A. J., Holtzman, D. A., Jackson, S. P., and Tjian, R. (1989). Synergistic activation by the glutamine-rich domains of human transcription factor Sp1. *Cell* **59**, 827–836.
11. Lania, L., Majello, B., and De Luca, P. (1997). Transcriptional regulation by the Sp family proteins. *Int. J. Biochem. Cell Biol.* **29**, 1313–1323.
12. Philipsen, S., and Suske, G. (1999). A tale of three fingers: The family of mammalian Sp/XKLF transcription factors. *Nucleic Acids Res.* **27**, 2991–3000.
13. Bouwman, P., and Philipsen, S. (2002). Regulation of the activity of Sp1-related transcription factors. *Mol. Cell. Endocrinol.* **195**, 27–38.
14. Black, A. R., Black, J. D., and Azizkhan-Clifford, J. (2001). Sp1 and Krüppel-like factor family of transcription factors in cell growth regulation and cancer. *J. Cell Physiol.* **188**, 143–160.
15. Pugh, B. F., and Tjian, R. (1990). Mechanism of transcriptional activation by Sp1: Evidence for coactivators. *Cell* **61**, 1187–1197.
16. Smale, S. T., Schmidt, M. C., Berk, A. J., and Baltimore, D. (1990). Transcriptional activation by Sp1 as directed through TATA or initiator: Specific requirement for mammalian transcription factor IID. *Proc. Natl. Acad. Sci. USA* **87**, 4509–4513.
17. Tanese, N., Pugh, B. F., and Tjian, R. (1991). Coactivators for a proline-rich activator purified from the multisubunit human TFIID complex. *Genes Dev.* **5**, 2212–2224.
18. Emami, K. H., Burke, T. W., and Smale, S. T. (1998). Sp1 activation of a TATA-less promoter requires a species-specific interaction involving transcription factor IID. *Nucleic Acids Res.* **26**, 839–846.
19. Gill, G., and Tjian, R. (1992). Eukaryotic coactivators associated with the TATA box binding protein. *Curr. Opin. Genet. Dev.* **2**, 236–242.
20. Hernandez, N. (1993). TBP, a universal eukaryotic transcription factor? *Genes Dev.* **7**, 1291–1308.
21. Gill, G., Pascal, E., Tseng, Z. H., and Tjian, R. (1994). A glutamine-rich hydrophobic patch in transcription factor Sp1 contacts the dTAF$_{II}$110 component of the *Drosophila* TFIID complex and mediates transcriptional activation. *Proc. Natl. Acad. Sci. USA* **91**, 192–196.
22. Hoey, T., Weinzierl, R. O., Gill, G., Chen, J. L., Dynlacht, B. D., and Tjian, R. (1993). Molecular cloning and functional analysis of *Drosophila* TAF110 reveal properties expected of coactivators. *Cell* **72**, 247–260.
23. Rojo-Niersbach, E., Furukawa, T., and Tanese, N. (1999). Genetic dissection of hTAF(II)130 defines a hydrophobic surface required for interaction with glutamine-rich activators. *J. Biol. Chem.* **274**, 33778–33784.
24. Saluja, D., Vassallo, M. F., and Tanese, N. (1998). Distinct subdomains of human TAF(II)130 are required for interactions with glutamine-rich transcriptional activators. *Mol. Cell. Biol.* **18**, 5734–5743.
25. Emili, A., Greenblatt, J., and Ingles, C. J. (1994). Species-specific interaction of the glutamine-rich activation domains of Sp1 with the TATA box-binding protein. *Mol. Cell. Biol.* **14**, 1582–1593.
26. Chiang, C.-M., and Roeder, R. G. (1995). Cloning of an intrinsic human TFIID subunit that interacts with multiple transcriptional activators. *Science* **267**, 531–536.
27. Shao, Z., Ruppert, S., and Robbins, P. D. (1995). The retinoblastoma-susceptibility gene product binds directly to the human TATA-binding protein-associated factor TAFII250. *Proc. Natl. Acad. Sci. USA* **92**, 3115–3119.
28. Ryu, S., and Tjian, R. (1999). Purification of transcription cofactor complex CRSP. *Proc. Natl. Acad. Sci. USA* **96**, 7137–7142.
29. Ryu, S., Zhou, S., Ladurner, A. G., and Tjian, R. (1999). The transcriptional cofactor complex CRSP is required for activity of the enhancer-binding protein Sp1. *Nature* **397**, 446–450.
30. Taatjes, D. J., Naar, A. M., Andel, F., Nogales, E., and Tjian, R. (2002). Structure, function, and activator-induced conformations of the CRSP coactivator. *Science* **295**, 1058–1062.

31. Sun, X., Zhang, Y., Cho, H., Rickert, P., Lees, E., Lane, W., and Reinberg, D. (1998). NAT, a human complex containing Srb polypeptides that functions as a negative regulator of activated transcription. *Mol. Cell* **2**, 213–222.
32. Kim, Y. J., Bjorklund, S., Li, Y., Sayre, M. H., and Kornberg, R. D. (1994). A multiprotein mediator of transcriptional activation and its interaction with the C-terminal repeat domain of RNA polymerase II. *Cell* **77**, 599–608.
33. Jiang, Y. W., Veschambre, P., Erdjument-Bromage, H., Tempst, P., Conaway, J. W., Conaway, R. C., and Kornberg, R. D. (1998). Mammalian mediator of transcriptional regulation and its possible role as an end-point of signal transduction pathways. *Proc. Natl. Acad. Sci. USA* **95**, 8538–8543.
34. Boyer, T. G., Martin, M. E., Lees, E., Ricciardi, R. P., and Berk, A. J. (1999). Mammalian Srb/Mediator complex is targeted by adenovirus E1A protein. *Nature* **399**, 276–279.
35. Näär, A. M., Beaurang, P. A., Zhou, S., Abraham, S., Solomon, W., and Tjian, R. (1999). Composite co-activator ARC mediates chromatin-directed transcriptional activation. *Nature* **398**, 828–832.
36. Ito, M., Yuan, C. X., Malik, S., Gu, W., Fondell, J. D., Yamamura, S., Fu, Z. Y., Zhang, X., Qin, J., and Roeder, R. G. (1999). Identity between TRAP and SMCC complexes indicates novel pathways for the function of nuclear receptors and diverse mammalian activators. *Mol. Cell* **3**, 361–370.
37. Rachez, C., Suldan, Z., Ward, J., Chang, C. P., Burakov, D., Erdjument-Bromage, H., Tempst, P., and Freedman, L. P. (1998). A novel protein complex that interacts with the vitamin D3 receptor in a ligand-dependent manner and enhances VDR transactivation in a cell-free system. *Genes Dev.* **12**, 1787–1800.
38. Roos, M. D., Su, K., Baker, J. R., and Kudlow, J. E. (1997). O-glycosylation of an Sp1-derived peptide blocks known Sp1 protein interactions. *Mol. Cell. Biol.* **17**, 6472–6480.
39. Yang, X., Su, K., Roos, M. D., Chang, Q., Paterson, A. J., and Kudlow, J. E. (2001). O-linkage of N-acetylglucosamine to Sp1 activation domain inhibits its transcriptional capability. *Proc. Natl. Acad. Sci. USA* **98**, 6611–6616.
40. Gottlieb, T. M., and Jackson, S. P. (1993). The DNA-dependent protein kinase: Requirement for DNA ends and association with Ku antigen. *Cell* **72**, 131–142.
41. Jackson, S. P., MacDonald, J. J., Lees-Miller, S., and Tjian, R. (1990). GC box binding induces phosphorylation of Sp1 by a DNA-dependent protein kinase. *Cell* **63**, 155–165.
42. Black, A. R., Jensen, D., Lin, S. Y., and Azizkhan, J. C. (1999). Growth/cell cycle regulation of Sp1 phosphorylation. *J. Biol. Chem.* **274**, 1207–1215.
43. Borellini, F., Aquino, A., Josephs, S. F., and Glazer, R. I. (1990). Increased expression and DNA-binding activity of transcription factor Sp1 in doxorubicin-resistant HL-60 leukemia cells. *Mol. Cell. Biol.* **10**, 5541–5547.
44. Han, I., and Kudlow, J. E. (1997). Reduced O-glycosylation of Sp1 is associated with increased proteasome susceptibility. *Mol. Cell. Biol.* **17**, 2550–2558.
45. Su, K., Roos, M. D., Yang, X., Han, I., Paterson, A. J., and Kudlow, J. E. (1999). An N-terminal region of Sp1 targets its proteasome-dependent degradation *in vitro*. *J. Biol. Chem.* **274**, 15194–15202.
46. Su, K., Yang, X., Roos, M. D., Paterson, A. J., and Kudlow, J. E. (2000). Human Sug1/p45 is involved in the proteasome-dependent degradation of Sp1. *Biochem. J.* **348 Pt. 2**, 281–289.
47. Dennig, J., Beato, M., and Suske, G. (1996). An inhibitor domain in Sp3 regulates its glutamine-rich activation domains. *EMBO J.* **15**, 5659–5667.
48. Braun, H., Koop, R., Ertmer, A., Nacht, S., and Suske, G. (2001). Transcription factor Sp3 is regulated by acetylation. *Nucleic Acids Res.* **29**, 4994–5000.

49. Marin, M., Karis, A., Visser, P., Grosveld, F., and Phillipsen, S. (1997). Transcription factor Sp1 is essential for early embryonic development but dispensable for cell growth and differentiation. *Cell* **89,** 619–628.
50. Gollner, H., Bouwman, P., Mangold, M., Karis, A., Braun, H., Rohner, I., Del Rey, A., Besedovsky, H. O., Meinhardt, A., van den Broek, M., Cutforth, T., Grosveld, F., Philipsen, S., and Suske, G. (2001). Complex phenotype of mice homozygous for a null mutation in the Sp4 transcription factor gene. *Genes Cells* **6,** 689–697.
51. Supp, D. M., Witte, D. P., Branford, W. W., Smith, E. P., and Potter, S. S. (1996). Sp4, a member of the Sp1-family of zinc finger transcription factors, is required for normal murine growth, viability, and male fertility. *Dev. Biol.* **176,** 284–299.
52. Nguyen-Tran, V. T., Kubalak, S. W., Minamisawa, S., Fiset, C., Wollert, K. C., Brown, A. B., Ruiz-Lozano, P., Barrere-Lemaire, S., Kondo, R., Norman, L. W., Gourdie, R. G., Rahme, M. M., Feld, G. K., Clark, R. B., Giles, W. R., and Chien, K. R. (2000). A novel genetic pathway for sudden cardiac death via defects in the transition between ventricular and conduction system cell lineages. *Cell* **102,** 671–682.
53. Harrison, S. M., Houzelstein, D., Dunwoodie, S. L., and Beddington, R. S. (2000). Sp5, a new member of the Sp1 family, is dynamically expressed during development and genetically interacts with Brachyury. *Dev. Biol.* **227,** 358–372.
54. Nakashima, K., Zhou, X., Kunkel, G., Zhang, Z., Deng, J. M., Behringer, R. R., and de Crombrugghe, B. (2002). The novel zinc finger-containing transcription factor osterix is required for osteoblast differentiation and bone formation. *Cell* **108,** 17–29.
55. Gollner, H., Dani, C., Phillips, B., Philipsen, S., and Suske, G. (2001). Impaired ossification in mice lacking the transcription factor Sp3. *Mech. Dev.* **106,** 77–83.
56. Khan, S., Abdelrahim, M., Samudio, I., and Safe, S. (2003). Estrogen receptor/Sp1 complexes are required for induction of *cad* gene expression by 17β-estradiol in breast cancer cells. *Endocrinology* **144,** 2325–2335.
57. Wang, W., Dong, L., Saville, B., and Safe, S. (1999). Transcriptional activation of E2F1 gene expression by 17β-estradiol in MCF-7 cells is regulated by NF-Y–Sp1/estrogen receptor interactions. *Mol. Endocrinol.* **13,** 1373–1387.
58. Ngwenya, S., and Safe, S. (2003). Cell context-dependent differences in the induction of E2F-1 gene expression by 17β-estradiol in MCF-7 and ZR 75 cells. *Endocrinology* **144,** 1675–1685.
59. Xiao, H., Hasegawa, T., and Isobe, K. (1999). Both Sp1 and Sp3 are responsible for p21^{wafl} promoter activity induced by histone deacetylase inhibitor in NIH3T3 cells. *J. Cell Biochem.* **73,** 291–302.
60. Xiao, H., Hasegawa, T., and Isobe, K.-I. (2000). p300 collaborates with Sp1 and Sp3 in p21$^{waf1/cip1}$ promoter activation induced by histone deacetylase inhibitor. *J. Biol. Chem.* **275,** 1371–1376.
61. Sowa, Y., Orita, T., Minamikawa, S., Nakano, K., Mizuno, T., Nomura, H., and Sakai, T. (1997). Histone deacetylase inhibitor activates the WAF1/Cip1 gene promoter through the Sp1 sites. *Biochem. Biophys. Res. Commun.* **241,** 142–150.
62. Nakano, K., Mizuno, T., Sowa, Y., Orita, T., Yoshino, T., Okuyama, Y., Fujita, T., Ohtani-Fujita, N., Matsukawa, Y., Tokino, T., Yamagishi, H., Oka, T., Nomura, H., and Sakai, T. (1997). Butyrate activates the WAF1/Cip1 gene promoter through Sp1 sites in a p53-negative human colon cancer cell line. *J. Biol. Chem.* **272,** 22199–22206.
63. Sowa, Y., Orita, T., Minamikawa-Hiranabe, S., Mizuno, T., Nomura, H., and Sakai, T. (1999). Sp3, but not Sp1, mediates the transcriptional activation of the p21/WAF1/Cip1 gene promoter by histone deacetylase inhibitor. *Cancer Res.* **59,** 4266–4270.
64. Pagliuca, A., Gallo, P., and Lania, L. (2000). Differential role for Sp1/Sp3 transcription factors in the regulation of the promoter activity of multiple cyclin-dependent kinase inhibitor genes. *J. Cell Biochem.* **76,** 360–367.

65. Koutsodontis, G., Tentes, I., Papakosta, P., Moustakas, A., and Kardassis, D. (2001). Sp1 plays a critical role in the transcriptional activation of the human cyclin-dependent kinase inhibitor p21(WAF1/Cip1) gene by the p53 tumor suppressor protein. *J. Biol. Chem.* **276,** 29116–29125.
66. Zhang, W., Geiman, D. E., Shields, J. M., Dang, D. T., Mahatan, C. S., Kaestner, K. H., Biggs, J. R., Kraft, A. S., and Yang, V. W. (2000). The gut-enriched Kruppel-like factor (Kruppel-like factor 4) mediates the transactivating effect of p53 on the p21WAF1/Cip1 promoter. *J. Biol. Chem.* **275,** 18391–18398.
67. Karlseder, J., Rotheneder, H., and Wintersberger, E. (1996). Interaction of Sp1 with the growth- and cell cycle-regulated transcription factor E2F. *Mol. Cell. Biol.* **16,** 1659–1667.
68. Lin, S.-Y., Black, A. R., Kostic, D., Pajovic, S., Hoover, C. N., and Azizkhan, J. C. (1996). Cell cycle-regulated association of E2F1 and Sp1 is related to their functional interaction. *Mol. Cell. Biol.* **16,** 1668–1675.
69. Wang, F., Wang, W., and Safe, S. (1999). Regulation of constitutive gene expression through interactions of Sp1 protein with the nuclear aryl hydrocarbon receptor complex. *Biochemistry* **38,** 11490–11500.
70. Kobayashi, A., Sogawa, K., and Fujii-Kuriyama, Y. (1996). Cooperative interaction between AhR·Arnt and Sp1 for the drug-inducible expression of *CYP1A1* gene. *J. Biol. Chem.* **271,** 12310–12316.
71. Santini, M. P., Talora, C., Seki, T., Bolgan, L., and Dotto, G. P. (2001). Cross talk among calcineurin, Sp1/Sp3, and NFAT in control of p21(WAF1/CIP1) expression in keratinocyte differentiation. *Proc. Natl. Acad. Sci. USA* **98,** 9575–9580.
72. Kardassis, D., Papakosta, P., Pardali, K., and Moustakas, A. (1999). c-Jun transactivates the promoter of the human p21$^{\text{WAF1/Cip1}}$ gene by acting as a superactivator of the ubiquitous transcription factor Sp1. *J. Biol. Chem.* **274,** 29572–29581.
73. Wang, C. H., Tsao, Y. P., Chen, H. J., Chen, H. L., Wang, H. W., and Chen, S. L. (2000). Transcriptional repression of p21 (Waf1/Cip1/Sdi1) gene by c-jun through Sp1 site. *Biochem. Biophys. Res. Commun.* **270,** 303–310.
74. Gartel, A. L., Ye, X., Goufman, E., Shianov, P., Hay, N., Najmabadi, F., and Tyner, A. L. (2001). Myc represses the p21(WAF1/CIP1) promoter and interacts with Sp1/Sp3. *Proc. Natl. Acad. Sci. USA* **98,** 4510–4515.
75. Moustakas, A., and Kardassis, D. (1998). Regulation of the human p21/WAF1/Cip1 promoter in hepatic cells by functional interactions between Sp1 and Smad family members. *Proc. Natl. Acad. Sci. USA* **95,** 6733–6738.
76. Pardali, K., Kurisaki, A., Moren, A., ten Dijke, P., Kardassis, D., and Moustakas, A. (2000). Role of Smad proteins and transcription factor Sp1 in p21(Waf1/Cip1) regulation by transforming growth factor-beta. *J. Biol. Chem.* **275,** 29244–29256.
77. Merika, M., and Orkin, S. H. (1995). Functional synergy and physical interactions of the erythroid transcription factor GATA-1 with the Krüppel family proteins Sp1 and EKLF. *Mol. Cell. Biol.* **15,** 2437–2447.
78. Roder, K., Wolf, S. S., Larkin, K. L., and Schweizer, M. (1999). Interaction between the two ubiquitously expressed transcription factors NF-Y and Sp1. *Gene* **234,** 61–69.
79. Yamada, K., Tanaka, T., Miyamoto, K., and Noguchi, T. (2000). Sp family members and nuclear factor-Y cooperatively stimulate transcription from the rat pyruvate kinase M gene distal promoter region via their direct interactions. *J. Biol. Chem.* **275,** 18129–18137.
80. Liang, F., Schaufele, F., and Gardner, D. G. (2001). Functional interaction of NF-Y and Sp1 is required for type a natriuretic peptide receptor gene transcription. *J. Biol. Chem.* **276,** 1516–1522.
81. Cohen, H. T., Zhou, M., Welsh, A. M., Zarghamee, S., Scholz, H., Mukhopadhyay, D., Kishida, T., Zbar, B., Knebelmann, B., and Sukhatme, V. P. (1999). An important von

Hippel-Lindau tumor suppressor domain mediates Sp1-binding and self-association. *Biochem. Biophys. Res. Commun.* **266,** 43–50.
82. Mukhopadhyay, D., Knebelmann, B., Cohen, H. T., Ananth, S., and Sukhatme, V. P. (1997). The von Hippel-Lindau tumor suppressor gene product interacts with Sp1 to repress vascular endothelial growth factor promoter activity. *Mol. Cell. Biol.* **17,** 5629–5639.
83. Biesiada, E., Hamamori, Y., Kedes, L., and Sartorelli, V. (1999). Myogenic basic helix–loop–helix proteins and Sp1 interact as components of a multiprotein transcriptional complex required for activity of the human cardiac α-actin promoter. *Mol. Cell. Biol.* **19,** 2577–2584.
84. Doetzlhofer, A., Rotheneder, H., Lagger, G., Koranda, M., Kurtev, V., Brosch, G., Wintersberger, E., and Seiser, C. (1999). Histone deacetylase 1 can repress transcription by binding to Sp1. *Mol. Cell. Biol.* **19,** 5504–5511.
85. Vallian, S., Chin, K. V., and Chang, K. S. (1998). The promyelocytic leukemia protein interacts with Sp1 and inhibits its transactivation of the epidermal growth factor receptor promoter. *Mol. Cell. Biol.* **18,** 7147–7156.
86. Ding, H., Benotmane, A. M., Suske, G., Collen, D., and Belayew, A. (1999). Functional interactions between Sp1 or Sp3 and the helicase-like transcription factor mediate basal expression from the human plasminogen activator inhibitor-1 gene. *J. Biol. Chem.* **274,** 19573–19580.
87. Lee, J. S., Galvin, K. M., and Shi, Y. (1993). Evidence for physical interaction between the zinc-finger transcription factors YY1 and Sp1. *Proc. Natl. Acad. Sci. USA* **90,** 6145–6149.
88. Johnson-Pais, T., Degnin, C., and Thayer, M. J. (2001). pRB induces Sp1 activity by relieving inhibition mediated by MDM2. *Proc. Natl. Acad. Sci. USA* **98,** 2211–2216.
89. Chen, B. K., and Chang, W. C. (2000). Functional interaction between c-Jun and promoter factor Sp1 in epidermal growth factor-induced gene expression of human 12(S)-lipoxygenase. *Proc. Natl. Acad. Sci. USA* **97,** 10406–10411.
90. Blaine, S. A., Wick, M., Dessev, C., and Nemenoff, R. A. (2001). Induction of cPLA2 in lung epithelial cells and non-small cell lung cancer is mediated by Sp1 and c-Jun. *J. Biol. Chem.* **276,** 42737–42743.
91. Melnikova, I. N., and Gardner, P. D. (2001). The signal transduction pathway underlying ion channel gene regulation by Sp1 C-Jun interactions. *J. Biol. Chem.* **276,** 19040–19045.
92. Pena, P., Reutens, A. T., Albanese, C., D'Amico, M., Watanabe, G., Donner, A., Shu, I. W., Williams, T., and Pestell, R. G. (1999). Activator protein-2 mediates transcriptional activation of the *CYP11A1* gene by interaction with Sp1 rather than binding to DNA. *Mol. Endocrinol.* **13,** 1402–1416.
93. Dunah, A. W., Jeong, H., Griffin, A., Kim, Y. M., Standaert, D. G., Hersch, S. M., Mouradian, M. M., Young, A. B., Tanese, N., and Krainc, D. (2002). Sp1 and TAFII130 transcriptional activity disrupted in early Huntington's disease. *Science* **296,** 2238–2243.
94. Li, S. H., Cheng, A. L., Zhou, H., Lam, S., Rao, M., Li, H., and Li, X. J. (2002). Interaction of Huntington disease protein with transcriptional activator Sp1. *Mol. Cell. Biol.* **22,** 1277–1287.
95. Haidweger, E., Novy, M., and Rotheneder, H. (2001). Modulation of Sp1 activity by a cyclin A/CDK complex. *J. Mol. Biol.* **306,** 201–212.
96. Strom, A. C., Forsberg, M., Lillhager, P., and Westin, G. (1996). The transcription factors Sp1 and Oct-1 interact physically to regulate human U2 snRNA gene expression. *Nucleic Acids Res.* **24,** 1981–1986.
97. Braun, H., and Suske, G. (1998). Combinatorial action of HNF3 and Sp family transcription factors in the activation of the rabbit uteroglobin/CC10 promoter. *J. Biol. Chem.* **273,** 9821–9828.
98. Kardassis, D., Falvey, E., Tsantili, P., Hadzopoulou-Cladaras, M., and Zannis, V. (2002). Direct physical interactions between HNF-4 and Sp1 mediate synergistic transactivation of the apolipoprotein CIII promoter. *Biochemistry* **41,** 1217–1228.

99. Chicas, A., Molina, P., and Bargonetti, J. (2000). Mutant p53 forms a complex with Sp1 on HIV-LTR DNA. *Biochem. Biophys. Res. Commun.* **279,** 383–390.
100. Borellini, F., and Glazer, R. I. (1993). Induction of Sp1-p53 DNA-binding heterocomplexes during granulocyte/macrophage colony-stimulating factor-dependent proliferation in human erythroleukemia cell line TF-1. *J. Biol. Chem.* **268,** 7923–7928.
101. Li, B., and Lee, M. Y. (2001). Transcriptional regulation of the human DNA polymerase δ catalytic subunit gene *POLD1* by p53 tumor suppressor and Sp1. *J. Biol. Chem.* **276,** 29729–29739.
102. Krainc, D., Bai, G., Okamoto, S., Carles, M., Kusiak, J. W., Brent, R. N., and Lipton, S. A. (1998). Synergistic activation of the *N*-methyl-D-aspartate receptor subunit 1 promoter by myocyte enhancer factor 2C and Sp1. *J. Biol. Chem.* **273,** 26218–26224.
103. Feng, X. H., Lin, X., and Derynck, R. (2000). Smad2, Smad3, and Smad4 cooperate with Sp1 to induce p15^{Ink4B} transcription in response to TGF-β. *EMBO J.* **19,** 5178–5193.
104. Botella, L. M., Sanchez-Elsner, T., Rius, C., Corbi, A., and Bernabeu, C. (2001). Identification of a critical Sp1 site within the endoglin promoter and its involvement in the transforming growth factor-β stimulation. *J. Biol. Chem.* **276,** 34486–34494.
105. Brodin, G., Ahgren, A., ten Dijke, P., Heldin, C. H., and Heuchel, R. (2000). Efficient TGF-β induction of the Smad7 gene requires cooperation between AP-1, Sp1, and Smad proteins on the mouse Smad7 promoter. *J. Biol. Chem.* **275,** 29023–29030.
106. Poncelet, A. C., and Schnaper, H. W. (2001). Sp1 and Smad proteins cooperate to mediate transforming growth factor-β1-induced α2(I) collagen expression in human glomerular mesangial cells. *J. Biol. Chem.* **276,** 6983–6992.
107. Zhang, W., Ou, J., Inagaki, Y., Greenwel, P., and Ramirez, F. (2000). Synergistic cooperation between Sp1 and Smad3/Smad4 mediates transforming growth factor β1 stimulation of α2(I)-collagen (COL1A2) transcription. *J. Biol. Chem.* **275,** 39237–39245.
108. Shetty, S., Takahashi, T., Matsui, H., Ayengar, R., and Raghow, R. (1999). Transcriptional autorepression of Msx1 gene is mediated by interactions of Msx1 protein with a multiprotein transcriptional complex containing TATA-binding protein, Sp1, and cAMP-response-element-binding protein-binding protein (CBP/p300). *Biochem. J.* **339,** 751–758.
109. Sif, S., and Gilmore, T. D. (1994). Interaction of the v-Rel oncoprotein with cellular transcription factor Sp1. *J. Virol.* **68,** 7131–7138.
110. Jeang, K.-T., Chun, R., Lin, N. H., Gatignol, A., Glabe, C. G., and Fan, H. (1993). *In vitro* and *in vivo* binding of human immunodeficiency virus type 1 Tat protein and Sp1 transcription factor. *J. Virol.* **67,** 6224–6233.
111. Li, R., Knight, J. D., Jackson, S. P., Tjian, R., and Botchan, M. R. (1991). Direct interaction between Sp1 and the BPV enhancer E2 protein mediates synergistic activation of transcription. *Cell* **65,** 493–505.
112. Kim, H. S., Lee, J. K., and Tsai, S. Y. (1995). E1a activation of insulin receptor gene expression is mediated by Sp1-binding sites. *Mol. Endocrinol.* **9,** 178–186.
113. Liu, F., and Green, M. R. (1994). Promoter targeting by adenovirus E1a through interaction with different cellular DNA-binding domains. *Nature* **368,** 520–525.
114. Kim, S.-J., Onwuta, U. S., Lee, Y. I., Li, R., Botchan, M. R., and Robbins, P. D. (1992). The retinoblastoma gene product regulates Sp1-mediated transcription. *Mol. Cell. Biol.* **12,** 2455–2463.
115. Chen, L. I., Nishinaka, T., Kwan, K., Kitabayashi, I., Yokoyama, K., Fu, Y. H., Grünwald, S., and Chiu, R. (1994). The retinoblastoma gene product RB stimulates Sp1-mediated transcription by liberating Sp1 from a negative regulator. *Mol. Cell. Biol.* **14,** 4380–4389.
116. Datta, P. K., Raychaudhuri, P., and Bagchi, S. (1995). Association of p107 with Sp1: Genetically separable regions of p107 are involved in regulation of E2F- and Sp1-dependent transcription. *Mol. Cell. Biol.* **15,** 5444–5452.

117. Song, J., Ugai, H., Kanazawa, I., Sun, K., and Yokoyama, K. K. (2001). Independent repression of a GC-rich housekeeping gene by Sp1 and MAZ involves the same cis-elements. *J. Biol. Chem.* **276,** 19897–19904.
118. Wieczorek, E., Lin, Z., Perkins, E. B., Law, D. J., Merchant, J. L., and Zehner, Z. E. (2000). The zinc finger repressor, ZBP-89, binds to the silencer element of the human vimentin gene and complexes with the transcriptional activator, Sp1. *J. Biol. Chem.* **275,** 12879–12888.
119. Tsai, M. J., and O'Malley, B. W. (1994). Molecular mechanisms of action of steroid/thyroid receptor superfamily members. *Annu. Rev. Biochem.* **63,** 451–486.
120. Mangelsdorf, D. J., Thummel, C., Beato, M., Herrlich, P., Schutz, G., Umesono, K., Blumberg, B., Kastner, P., Mark, M., Chambon, P., and Evans, R. M. (1995). The nuclear receptor superfamily: The second decade. *Cell* **83,** 835–839.
121. Beato, M., Herrlich, P., and Schutz, G. (1995). Steroid hormone receptors: Many actors in search of a plot. *Cell* **83,** 851–857.
122. Olefsky, J. M. (2001). Nuclear receptor minireview series. *J. Biol. Chem.* **276,** 36863–36864.
123. Altucci, L., and Gronemeyer, H. (2001). Nuclear receptors in cell life and death. *Trends Endocrinol. Metab.* **12,** 460–468.
124. Kuiper, G. G., Enmark, E., Pelto-Huikko, M., Nilsson, S., and Gustafsson, J. A. (1996). Cloning of a novel receptor expressed in rat prostate and ovary. *Proc. Natl. Acad. Sci. USA* **93,** 5925–5930.
125. Mosselman, S., Polman, J., and Dijkema, R. (1996). ER_β: Identification and characterization of a novel human estrogen receptor. *FEBS Lett.* **392,** 49–53.
126. Enmark, E., Pelto-Huikko, M., Grandien, K., Lagercrantz, S., Lagercrantz, J., Fried, G., Nordenskjold, M., and Gustafsson, J. A. (1997). Human estrogen receptor β-gene structure, chromosomal localization, and expression pattern. *J. Clin. Endocrinol. Metab.* **82,** 4258–4265.
127. Tremblay, G. B., Tremblay, A., Copeland, N. G., Gilbert, D. J., Jenkins, N. A., Labrie, F., and Giguère, V. (1997). Cloning, chromosomal localization, and functional analysis of the murine estrogen receptor β. *Mol. Endocrinol.* **11,** 353–365.
128. Cowley, S. M., Hoare, S., Mosselman, S., and Parker, M. G. (1997). Estrogen receptors α and β form heterodimers on DNA. *J. Biol. Chem.* **272,** 19858–19862.
129. Dubik, D., and Shiu, R. P. C. (1992). Mechanism of estrogen activation of c-myc oncogene expression. *Oncogene* **7,** 1587–1594.
130. Wu-Peng, X. S., Pugliese, T. E., Dickerson, H. W., and Pentecost, B. T. (1992). Delineation of sites mediating estrogen regulation of the rat creatine kinase B gene. *Mol. Endocrinol.* **6,** 231–240.
131. Cavailles, V., Augereau, P., and Rochefort, H. (1993). Cathepsin D gene is controlled by a mixed promoter, and estrogens stimulate only TATA-dependent transcription. *Proc. Natl. Acad. Sci. USA* **90,** 203–207.
132. Augereau, P., Miralles, F., Cavailles, V., Gaudelet, C., Parker, M., and Rochefort, H. (1994). Characterization of the proximal estrogen-responsive element of human cathepsin D gene. *Mol. Endocrinol.* **8,** 693–703.
133. Krishnan, V., Wang, X., and Safe, S. (1994). Estrogen receptor-Sp1 complexes mediate estrogen-induced cathepsin D gene expression in MCF-7 human breast cancer cells. *J. Biol. Chem.* **269,** 15912–15917.
134. Krishnan, V., Porter, W., Santostefano, M., Wang, X., and Safe, S. (1995). Molecular mechanism of inhibition of estrogen-induced cathepsin D gene expression by 2,3,7,8-tetrachlorodibenzo-p-dioxin (TCDD) in MCF-7 cells. *Mol. Cell. Biol.* **15,** 6710–6719.
135. Vyhlidal, C., Samudio, I., Kladde, M., and Safe, S. (2000). Transcriptional activation of transforming growth factor α by estradiol: Requirement for both a GC-rich site and an estrogen response element half-site. *J. Mol. Endocrinol.* **24,** 329–338.

136. Porter, W., Wang, F., Wang, W., Duan, R., and Safe, S. (1996). Role of estrogen receptor/Sp1 complexes in estrogen-induced heat shock protein 27 gene expression. *Mol. Endocrinol.* **10,** 1371–1378.
137. Petz, L. N., and Nardulli, A. M. (2000). Sp1 binding sites and an estrogen response element half-site are involved in regulation of the human progesterone receptor A promoter. *Mol. Endocrinol.* **14,** 972–985.
138. Dennig, J., Hagen, G., Beato, M., and Suske, G. (1995). Members of the Sp transcription factor family control transcription from the uteroglobin promoter. *J. Biol. Chem.* **270,** 12737–12744.
139. Suske, G., Wenz, M., Cato, A. C., and Beato, M. (1983). The uteroglobin gene region: Hormonal regulation, repetitive elements, and complete nucleotide sequence of the gene. *Nucleic Acids Res.* **11,** 2257–2271.
140. Scholz, A., Truss, M., and Beato, M. (1998). Hormone-induced recruitment of Sp1 mediates estrogen activation of the rabbit uteroglobin gene in endometrial epithelium. *J. Biol. Chem.* **273,** 4360–4366.
141. Batistuzzo de Medeiros, S. R., Krey, G., Hihi, A. K., and Wahli, W. (1997). Functional interactions between the estrogen receptor and the transcription activator Sp1 regulate the estrogen-dependent transcriptional activity of the vitellogenin A1 *io* promoter. *J. Biol. Chem.* **272,** 18250–18260.
142. Weisz, A., and Rosales, R. (1990). Identification of an estrogen response element upstream of the human c-*fos* gene that binds the estrogen receptor and the AP-1 transcription factor. *Nucleic Acids Res.* **18,** 5097–5106.
143. Duan, R., Porter, W., and Safe, S. (1998). Estrogen-induced c-*fos* protooncogene expression in MCF-7 human breast cancer cells: Role of estrogen receptor Sp1 complex formation. *Endocrinology* **139,** 1981–1990.
144. Porter, W., Saville, B., Hoivik, D., and Safe, S. (1997). Functional synergy between the transcription factor Sp1 and the estrogen receptor. *Mol. Endocrinol.* **11,** 1569–1580.
145. Matthews, M. A., Markowitz, R. B., and Dynan, W. S. (1992). *In vitro* activation of transcription by the human T-cell leukemia virus type I Tax protein. *Mol. Cell. Biol.* **12,** 1986–1996.
146. Zhao, L. J., and Giam, C. Z. (1992). Human T-cell lymphotropic virus type I (HTLV-I) transcriptional activator, Tax, enhances CREB binding to HTLV-I 21-base-pair repeats by protein–protein interaction. *Proc. Natl. Acad. Sci. USA* **89,** 7070–7074.
147. Armstrong, A. P., Franklin, A. A., Uittenbogaard, M. N., Giebler, H. A., and Nyborg, J. K. (1993). Pleiotropic effect of the human T-cell leukemia virus Tax protein on the DNA binding activity of eukaryotic transcription factors. *Proc. Natl. Acad. Sci. USA* **90,** 7303–7307.
148. Franklin, A. A., Kubik, M. F., Uittenbogaard, M. N., Brauweiler, A., Utaisincharoen, P., Matthews, M. A., Dynan, W. S., Hoeffler, J. P., and Nyborg, J. K. (1993). Transactivation by the human T-cell leukemia virus Tax protein is mediated through enhanced binding of activating transcription factor-2 (ATF-2) ATF-2 response and cAMP element-binding protein (CREB). *J. Biol. Chem.* **268,** 21225–21231.
149. Anderson, M. G., and Dynan, W. S. (1994). Quantitative studies of the effect of HTLV-I Tax protein on CREB protein-DNA binding. *Nucleic Acids Res.* **22,** 3194–3201.
150. Sanchez, H. B., Yieh, L., and Osborne, T. F. (1995). Cooperation by sterol regulatory element-binding protein and Sp1 in sterol regulation of low density lipoprotein receptor gene. *J. Biol. Chem.* **270,** 1161–1169.
151. Zwijsen, R. M., Wientjens, E., Klompmaker, R., van der Sman, J., Bernards, R., and Michalides, R. J. (1997). CDK-independent activation of estrogen receptor by cyclin D1. *Cell* **88,** 405–415.

152. Saville, B., Wormke, M., Wang, F., Nguyen, T., Enmark, E., Kuiper, G., Gustafsson, J.-A., and Safe, S. (2000). Ligand-, cell-, and estrogen receptor subtype (α/β)-dependent activation at GC-rich (Sp1) promoter elements. *J. Biol. Chem.* **275,** 5379–5387.
153. Berry, M., Nunez, A.-M., and Chambon, P. (1989). Estrogen-responsive element of the human pS2 gene is an imperfectly palindromic sequence. *Proc. Natl. Acad. Sci. USA* **86,** 1218–1222.
154. Zacharewski, T. R., Bondy, K. L., McDonell, P., and Wu, Z. F. (1994). Antiestrogenic effects of 2,3,7,8-tetrachlorodibenzo-*p*-dioxin on 17β-estradiol-induced pS2 expression. *Cancer Res.* **54,** 2707–2713.
155. Wang, F., Hoivik, D., Pollenz, R., and Safe, S. (1998). Functional and physical interactions between the estrogen receptor-Sp1 and the nuclear aryl hydrocarbon receptor complexes. *Nucleic Acids Res.* **26,** 3044–3052.
156. Xie, W., Duan, R., and Safe, S. (1999). Estrogen induces adenosine deaminase gene expression in MCF-7 human breast cancer cells: Role of estrogen receptor-Sp1 interactions. *Endocrinology* **140,** 219–227.
157. Sun, G., Porter, W., and Safe, S. (1998). Estrogen-induced retinoic acid receptor α1 gene expression: Role of estrogen receptor-Sp1 complex. *Mol. Endocrinol.* **12,** 882–890.
158. Qin, C., Singh, P., and Safe, S. (1999). Transcriptional activation of insulin-like growth factor binding protein 4 by 17β-estradiol in MCF-7 cells: Role of estrogen receptor-Sp1 complexes. *Endocrinology* **140,** 2501–2508.
159. Dong, L., Wang, W., Wang, F., Stoner, M., Reed, J. C., Harigai, M., Kladde, M., Vyhlidal, C., and Safe, S. (1999). Mechanisms of transcriptional activation of *bcl-2* gene expression by 17β-estradiol in breast cancer cells. *J. Biol. Chem.* **174,** 32099–32107.
160. Xie, W., Duan, R., Chen, I., Samudio, I., and Safe, S. (2000). Transcriptional activation of thymidylate synthase by 17β-estradiol in MCF-7 human breast cancer cells. *Endocrinology* **141,** 2439–2449.
161. Samudio, I., Vyhlidal, C., Wang, F., Stoner, M., Chen, I., Kladde, M., Barhoumi, R., Burghardt, R., and Safe, S. (2001). Transcriptional activation of DNA polymerase α gene expression in MCF-7 cells by 17β-estradiol. *Endocrinology* **142,** 1000–1008.
162. Castro-Rivera, E., Samudio, I., and Safe, S. (2001). Estrogen regulation of cyclin D1 gene expression in ZR-75 breast cancer cells involves multiple enhancer elements *J. Biol. Chem.* **276,** 30853–30861.
163. Wang, F., Samudio, I., and Safe, S. (2002). Transcriptional activation of rat creatine kinase B by 17β-estradiol in MCF-7 cells involves an estrogen responsive element and GC-rich sites. *J. Cell Biochem.* **84,** 156–172.
164. Salvatori, L., Ravenna, L., Felli, M. P., Cardillo, M. R., Russo, M. A., Frati, L., Gulino, A., and Petrangeli, E. (2000). Identification of an estrogen-mediated deoxyribonucleic acid-binding independent transactivation pathway on the epidermal growth factor receptor gene promoter. *Endocrinology* **141,** 2266–2274.
165. Tanaka, N., Yonekura, H., Yamagishi, S., Fujimori, H., Yamamoto, Y., and Yamamoto, H. (2000). The receptor for advanced glycation end products is induced by the glycation products themselves and tumor necrosis factor-α through nuclear factor-κB, and by 17β-estradiol through Sp-1 in human vascular endothelial cells. *J. Biol. Chem.* **275,** 25781–25790.
166. Li, C., Briggs, M. R., Ahlborn, T. E., Kraemer, F. B., and Liu, J. (2001). Requirement of Sp1 and estrogen receptor α interaction in 17β-estradiol-mediated transcriptional activation of the low density lipoprotein receptor gene expression. *Endocrinology* **142,** 1546–1553.
167. Jacobson, D., Pribnow, D., Herson, P. S., Maylie, J., and Adelman, J. P. (2003). Determinants contributing to estrogen-regulated expression of SK3. *Biochem. Biophys. Res. Commun.* **303,** 660–668.

168. Schultz, J. R., Petz, L. N., and Nardulli, A. M. (2003). Estrogen receptor α and Sp1 regulate progesterone receptor gene expression. *Mol. Cell. Endocrinol.* **201**, 165–175.
169. Duan, R., Xie, W., Burghardt, R., and Safe, S. (2001). Estrogen receptor-mediated activation of the serum response element in MCF-7 cells through MAPK-dependent phosphorylation of Elk-1. *J. Biol. Chem.* **276**, 11590–11598.
170. Xing, W., and Archer, T. K. (1998). Upstream stimulatory factors mediate estrogen receptor activation of the cathepsin D promoter. *Mol. Endocrinol.* **12**, 1310–1321.
171. Abdelrahim, M., Samudio, I., Smith, R., Burghardt, R., and Safe, S. (2002). Small inhibitory RNA duplexes for Sp1 mRNA block basal and estrogen-induced gene expression and cell cycle progression in MCF-7 breast cancer cells. *J. Biol. Chem.* **277**, 28815–28822.
172. Martini, P. G., and Katzenellenbogen, B. S. (2001). Regulation of prothymosin α gene expression by estrogen in estrogen receptor-containing breast cancer cells via upstream half-palindromic estrogen response element motifs. *Endocrinology* **142**, 3493–3501.
173. Paech, K., Webb, P., Kuiper, G. G., Nilsson, S., Gustafsson, J., Kushner, P. J., and Scanlan, T. S. (1997). Differential ligand activation of estrogen receptors ERα and ERβ at AP1 sites. *Science* **277**, 1508–1510.
174. Webb, P., Nguyen, P., Valentine, C., Lopez, G. N., Kwok, G. R., McInerney, E., Katzenellenbogen, B. S., Enmark, E., Gustafsson, J.-Å, Nilsson, S., and Kushner, P. J. (1999). The estrogen receptor enhances AP-1 activity by two distinct mechanisms with different requirements for receptor transactivation functions. *Mol. Endocrinol.* **13**, 1672–1685.
175. Webb, P., Lopez, G. N., Uht, R. M., and Kushner, P. J. (1995). Tamoxifen activation of the estrogen receptor/AP-1 pathway: Potential origin for the cell-specific estrogen-like effects of antiestrogens. *Mol. Endocrinol.* **9**, 443–456.
176. Uht, R. M., Anderson, C. M., Webb, P., and Kushner, P. J. (1997). Transcriptional activities of estrogen and glucocorticoid receptors are functionally integrated at the AP-1 response element. *Endocrinology* **138**, 2900–2908.
177. Weatherman, R. V., and Scanlan, T. S. (2001). Unique protein determinants of the subtype-selective ligand responses of the estrogen receptors (ERα and ERβ) at AP-2 sites. *J. Biol. Chem.* **276**, 3827–3832.
178. Kim, K., Thu, N., Saville, B., and Safe, S. (2003). Domains of estrogen receptor α (ERα) required for ERα/Sp1-mediated activation of GC-rich promoters by estrogens and antiestrogens in breast cancer cells. *Mol. Endocrinol.* **17**, 804–817.
179. Ruohola, J. K., Valve, E. M., Karkkainen, M. J., Joukov, V., Alitalo, K., and Härkönen, P. L. (1999). Vascular endothelial growth factors are differentially regulated by steroid hormones and antiestrogens in breast cancer cells. *Mol. Cell. Endocrinol.* **149**, 29–40.
180. Cullinan-Bove, K., and Koos, R. D. (1993). Vascular endothelial growth factor/vascular permeability factor expression in the rat uterus: Rapid stimulation by estrogen correlates with estrogen-induced increases in uterine capillary permeability and growth. *Endocrinology* **133**, 829–837.
181. Hyder, S. M., Stancel, G. M., Chiappetta, C., Murthy, L., Boettger-Tong, H. L., and Makela, S. (1996). Uterine expression of vascular endothelial growth factor is increased by estradiol and tamoxifen. *Cancer Res.* **56**, 3954–3960.
182. Mueller, M. D., Vigne, J. L., Minchenko, A., Lebovic, D. I., Leitman, D. C., and Taylor, R. N. (2000). Regulation of vascular endothelial growth factor (VEGF) gene transcription by estrogen receptors α and β. *Proc. Natl. Acad. Sci. USA* **97**, 10972–10977.
183. Stoner, M., Wang, F., Wormke, M., Nguyen, T., Samudio, I., Vyhlidal, C., Marme, D., Finkenzeller, G., and Safe, S. (2000). Inhibition of vascular endothelial growth factor expression in HEC1A endometrial cancer cells through interactions of estrogen receptor α and Sp3 proteins. *J. Biol. Chem.* **275**, 22769–22779.

184. Stoner, M., Wormke, M., Saville, B., Samudio, I., Qin, C., Abdelrahim, M., and Safe, S. (2004). Estrogen regulation of vascular endothelial growth factor gene expression in ZR-75 breast cancer cells through interaction of estrogen receptor α and Sp proteins. *Oncogene* **23**, 1052–1063.
185. Klinge, C. M. (2000). Estrogen receptor interactions with co-activators and corepressors. *Steroids* **65**, 227–251.
186. Chen, J. D. (2000). Steroid/nuclear receptor coactivators. *Vitam. Horm.* **58**, 391–448.
187. Glass, C. K., Rose, D. W., and Rosenfeld, M. G. (1997). Nuclear receptor coactivators. *Curr. Opin. Cell Biol.* **9**, 222–232.
188. Edwards, D. P. (1999). Coregulatory proteins in nuclear hormone receptor action. *Vitam. Horm.* **55**, 165–218.
189. McKenna, N. J., Xu, J., Nawaz, Z., Tsai, S. Y., Tsai, M. J., and O'Malley, B. W. (1999). Nuclear receptor coactivators: Multiple enzymes, multiple complexes, multiple functions. *J. Steroid Biochem. Mol. Biol.* **69**, 3–12.
190. Robyr, D., Wolffe, A. P., and Wahli, W. (2000). Nuclear hormone receptor coregulators in action: Diversity for shared tasks. *Mol. Endocrinol.* **14**, 329–347.
191. Lemon, B. D., and Freedman, L. P. (1999). Nuclear receptor cofactors as chromatin remodelers. *Curr. Opin. Genet. Dev.* **9**, 499–504.
192. Hermanson, O., Glass, C. K., and Rosenfeld, M. G. (2002). Nuclear receptor coregulators: Multiple modes of modification. *Trends Endocrinol. Metab.* **13**, 55–60.
193. Owen, G. I., Richer, J. K., Tung, L., Takimoto, G., and Horwitz, K. B. (1998). Progesterone regulates transcription of the p21^{WAF1} cyclin-dependent kinase inhibitor gene through Sp1 and CBP/p300. *J. Biol. Chem.* **273**, 10696–10701.
194. Simmen, R. C. M., Chung, T. E., Imataka, H., Michel, F. J., Badinga, L., and Simmen, F. A. (1999). *Trans*-activation functions of the Sp-related nuclear factor, basic transcription element-binding protein, and progesterone receptor in endometrial epithelial cells. *Endocrinology* **140**, 2517–2525.
195. Lu, S., Jenster, G., and Epner, D. E. (2000). Androgen induction of cyclin-dependent kinase inhibitor p21 gene: Role of androgen receptor and transcription factor Sp1 complex. *Mol. Endocrinol.* **14**, 753–760.
196. Curtin, D., Jenkins, S., Farmer, N., Anderson, A. C., Haisenleder, D. J., Rissman, E., Wilson, E. M., and Shupnik, M. A. (2001). Androgen suppression of GnRH-stimulated rat LHβ gene transcription occurs through Sp1 sites in the distal GnRH-responsive promoter region. *Mol. Endocrinol.* **15**, 1906–1917.
197. Husmann, M., Dragneva, Y., Romahn, E., and Jehnichen, P. (2000). Nuclear receptors modulate the interaction of Sp1 and GC-rich DNA via ternary complex formation. *Biochem. J.* **352**, 763–772.
198. Suzuki, Y., Shimada, J., Shudo, K., Matsumura, M., Crippa, M. P., and Kojima, S. (1999). Physical interactions between retinoic acid receptor and Sp1: Mechanism for induction of urokinase by retinoic acid. *Blood* **93**, 4264–4276.
199. Shimada, J., Suzuki, Y., Kim, S. J., Wang, P. C., Matsumura, M., and Kojima, S. (2001). Transactivation via RAR/RXR-Sp1 interaction: Characterization of binding between Sp1 and GC box motif. *Mol. Endocrinol.* **15**, 1677–1692.
200. Sugawara, A., Uruno, A., Kudo, M., Ikeda, Y., Sato, K., Taniyama, Y., Ito, S., and Takeuchi, K. (2002). Transcription suppression of thromboxane receptor gene by peroxisome proliferator-activated receptor-γ via an interaction with Sp1 in vascular smooth muscle cells. *J. Biol. Chem.* **277**, 9676–9683.
201. Rohr, O., Aunis, D., and Schaeffer, E. (1997). COUP-TF and Sp1 interact and cooperate in the transcriptional activation of the human immunodeficiency virus type 1 long terminal repeat in human microglial cells. *J. Biol. Chem.* **272**, 31149–31155.

202. Pipaón, C., Tsai, S. Y., and Tsai, M. J. (1999). COUP-TF upregulates *NGFI-A* gene expression through an Sp1 binding site. *Mol. Cell. Biol.* **19,** 2734–2745.
203. Liu, Z., and Simpson, E. R. (1999). Molecular mechanism for cooperation between Sp1 and steroidogenic factor-1 (SF-1) to regulate bovine CYP11A gene expression. *Mol. Cell. Endocrinol.* **153,** 183–196.
204. Sugawara, T., Saito, M., and Fujimoto, S. (2000). Sp1 and SF-1 interact and cooperate in the regulation of human steroidogenic acute regulatory protein gene expression. *Endocrinology* **141,** 2895–2903.

Site-Specific DNA Damage Recognition by Enzyme-Induced Base Flipping

JAMES T. STIVERS

Department of Pharmacology and Molecular Sciences, The Johns Hopkins University, Baltimore, MD 21205

I. Why Do Enzymes Flip Bases?	39
II. Nonenzymatic Base Pair Breathing	41
A. NMR Studies	41
B. Computational Studies	43
III. Enzymatic Base Flipping	43
A. Structural Studies	44
B. Methods and Mechanism	44
IV. New Experimental Approaches	54
A. Preorganized Substrates	54
B. Nonpolar Damaged Base Isosteres	56
C. Damaged Base Pair Stability and Specific Recognition	58
V. Future Directions	60
References	61

One of the most interesting enigmas in enzymatic DNA recognition is the mechanism by which damaged bases buried in the DNA base stack are detected and subsequently excised. The essential enzyme combatants in this never-ending battle against genomic instability, the DNA repair glycosylases (Fig. 1A), must act specifically against a variety of base lesions that arise from oxidative reactions, deamination events, and alkylating agents, without also mistakenly removing normal nucleobases (1). This is a high stakes game: the cost of being slow is irreversible damage to the coding content of the genome, and the cost of being sloppy is the indiscriminate introduction of toxic abasic sites in DNA (2–4). Because many damaged bases are nonperturbing to the duplex structure and may differ from normal bases by only a single atom change, an extraordinary and highly conserved recognition mechanism has evolved to accomplish this task. This mechanism has been termed "base flipping," and involves the dramatic rotation of an entire damaged base and sugar from the DNA duplex using binding forces imposed by the enzyme (Fig. 1B) (5–7).

The problem facing these enzymes is both enormous and unique. Consider that a damaged base may compose only one out of 10^7 base pairs in the human

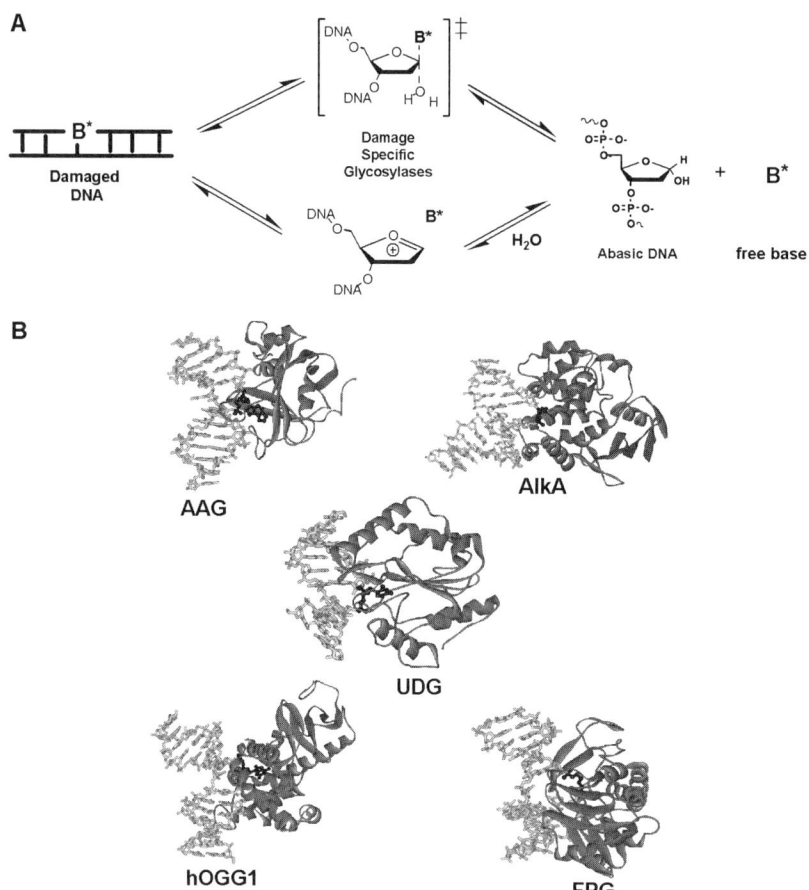

FIG. 1. The general reaction catalyzed by DNA glycosylases and the structures of several glycosylase-damaged DNA complexes. (A) DNA glycosylases recognize damaged bases (B°) in the vast context of genomic DNA and cleave the glycosidic bond, resulting in the release of the damaged base and the generation of an abasic site in DNA. Two general types of glycosylases exist: those that use water as the intial nucleophile (shown) and those that employ an active site amine (56). Two different chemical mechanisms are depicted, one involving concerted attack of water and displacement of the base, and one involving stepwise formation of a glycosyl cation intermediate followed by water attack. (B) Structures of five glycosylases that flip damaged bases into their active sites. The human (AAG) and bacterial (AlkA) alkyl adenine DNA glycosylases use a water nucleophile to remove alkylated purine bases from DNA (13, 35, 86, 87). The human (hOGG1) and bacterial (FPG) 8-oxoguanine DNA glycosylases remove oxidized guanine residues from 8-oxoG/C base pairs. Both of these enzymes employ amine nucleophiles (56). The remaining enzyme uracil DNA glycosylase (UDG) removes uracil from single-stranded DNA or from duplex DNA in the context of U/A or U/G base pairs. The base flipping mechanism of UDG has been subjected to extensive structural and kinetic analysis (10, 30–32, 64).

genome (*8, 9*). Thus, a crude estimate of the specificity for excision of a specific damaged base over a normal base of $(k_{cat}/K_m)^{specific}/(k_{cat}/K_m)^{nonspecific} \geq 10^7$ may be assumed. Although such issues as increased accessibility of damaged vs normal bases in the densely packed chromatin may account for some of this apparently large specificity, it is difficult to dismiss the magnitude of the discrimination achieved by the recognition mechanisms of these enzymes. Equally remarkable is that specificity is achieved without forming sequence specific contacts with the DNA substrate. Indeed, any specificity mechanism that arose during natural selection involving sequence specific interactions would have been selected against, because this mechanism compromises the ability of the organism to execute unbiased repair of DNA lesions regardless of the sequence context. Accordingly, DNA glycosylases present a unique evolutionary solution to site recognition that is fundamentally different from that of restriction enzymes, repressor proteins, or transcription factors.

Structural and other biophysical studies over the last four years have begun to reveal the basis for DNA glycosylase specificity and the critical role of DNA base flipping (*5, 10–16*). The system that has advanced most rapidly in this time span is that of the DNA repair enzyme uracil DNA glycosylase. This enzyme removes unwanted uracil residues in DNA that have arisen by spontaneous deamination of cytosine, or by misincorporation of dUTP during DNA replication (*17*). UDG has provided an experimentally tractable system to investigate the multistep process of base flipping, using a wide range of experimental approaches. The findings in this system should have widespread applicability given the highly conserved structural features of enzymes that flip bases (Fig. 1B). The general questions that have driven our research in base flipping over the last few years also direct the course of this chapter. To begin, we ask why has base flipping evolved as the sole mechanism for recognition and catalysis by DNA glycosylases? Second, what are the natures of the energetic barriers that an enzyme must overcome to extricate a base from its position in the DNA duplex? And finally, what temporal events occur during enzymatic base flipping, and how do the various steps contribute to catalytic specificity? We are continuing to develop new experimental tools to investigate base flipping, and we present these new approaches in the hope that they become generally useful in the investigation of other enzymes that flip bases.

I. Why Do Enzymes Flip Bases?

It is remarkable that all DNA glycosylases form structurally similar flipped-out complexes with their respective damaged bases, even though many of these enzymes share no sequence or structural similarity (Fig. 1B) (*10, 13, 15, 18–20*). This observation suggests that there exists a fundamental and shared biological problem for recognition of all damaged bases that can only be resolved by ejecting it from its position in the DNA duplex and positioning it in the confines of the enzyme active site. One driving force for this unified solution is the chemical requirement for access to the proton acceptor groups on the damaged base. The powerful catalytic benefit of base protonation is well documented for the nonenzymatic hydrolysis reactions of purine and pyrimidine bases, and serves to make the base electron deficient, thereby promoting flow of electrons away from the glycosidic bond (Fig. 2) (21–23). Without such interactions, it is difficult to imagine how these enzymes could begin to destabilize the N-glycosidic linkage as a prerequisite for bond cleavage. The second driving force derives from the constraint that these enzymes not form a large sequence specific interaction surface with the DNA (see preceding text). Accordingly, these enzymes were forced to evolve specific hydrogen bonding and aromatic stacking interactions with the damaged base, and sequence nonspecific interactions with the DNA backbone, that in their entirety provide enough favorable binding energy to disrupt the stacking and hydrogen

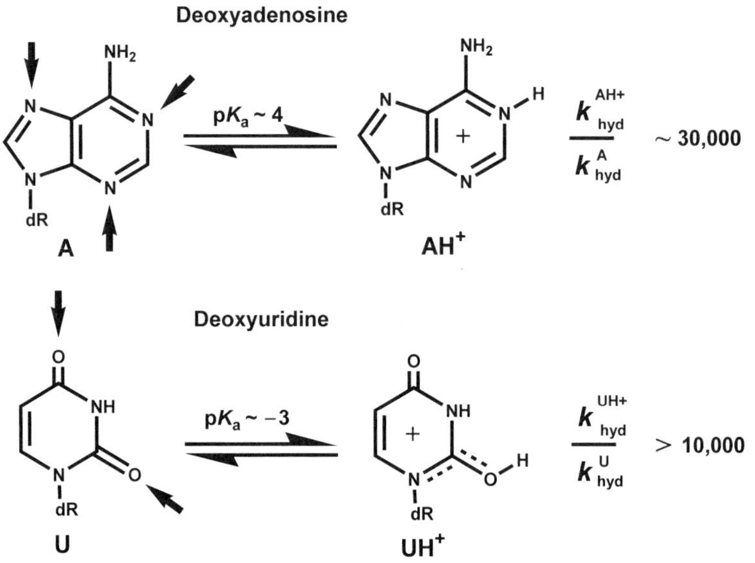

FIG. 2. The catalytic benefits of gaining access to the proton acceptor groups on purine and pyrimidine bases by base flipping. The nonenzymatic hydrolysis of protonated purine and pyrimidine nucleosides (k^{AH+} and k^{UH+}) is enhanced tremendously, because protonation of the leaving group is a powerful mechanism for making the base electron deficient (56). Enzymes would gain the same catalytic benefit by gaining access to these groups, which would provide a strong selection mechanism for evolution of a base flipping strategy.

bonding interactions of the damaged base in its duplex environment. Included in this energetic accounting is the free energy cost for DNA bending (see following text), which is a universal feature of all enzymes that flip bases. Thus, enzymes flip bases because the chemistry of glycosidic bond cleavage requires it, and flipping a base into an enzyme active site likely provides the only available means to access a large enough specific interaction surface to energetically drive this difficult transformation.

II. Nonenzymatic Base Pair Breathing

One rational place to begin understanding the process of enzymatic base flipping is to consider the process of spontaneous base pair opening (or breathing) in DNA (Fig. 3) (24). Although the enzymatic and nonenzymatic processes could involve different structural transformations in the DNA, studies of the spontaneous reaction mechanism reveal the *lowest energy pathway* for expelling a base from the native DNA duplex. Accordingly, uncovering the nature of the kinetic, thermodynamic, and structural barriers that the enzyme must overcome to enhance the efficiency of base pair opening is a useful reason to study the nonenzymatic reaction. These aspects of base pair opening have been investigated extensively using nuclear magnetic resonance spectroscopy (NMR) and computational approaches (25, 26), and the major findings relevant to enzymatic base flipping are summarized here.

A. NMR Studies

The primary experimental approach to study nonenzymatic base pair opening is to perform imino proton exchange experiments using NMR spectroscopy (25). Such experiments usually involve selectively perturbing the

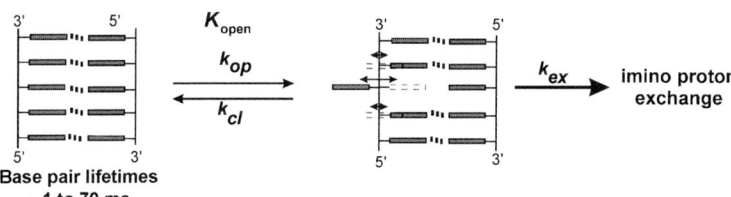

FIG. 3. Nonenzymatic base pair breathing. Normal base pairs in duplex DNA are in a dynamic equilibrium between a stacked and hydrogen bonded state in the duplex base stack and a solvent exposed state that is detectable by imino proton exchange. This process may be described by an equilibrium constant for base pair opening (K_{open}), and the rate constants for opening and closing (k_{op} and k_{cl}) and for exchange (k_{ex}). The rate-limiting step for exchange may be manipulated by the addition of base catalysts, which increase k_{ex} without altering k_{op} and k_{cl} (25).

magnetization in bulk water and then observing the time-dependent exchange of "magnetically labeled" water protons with the imino protons of the DNA bases. Under appropriate conditions, the exchange process is dependent on the opening rate of the DNA base pairs (k_{op}, Fig. 3), and thereby provides a useful probe to investigate the base pair dynamics of DNA in the millisecond to 100 ms time-regime. These studies have generally found that AT base pairs have shorter lifetimes ($1/k_{op} = 1-5$ ms) than do GC pairs ($1/k_{op} = 10-50$ ms), but that the DNA sequence context can also affect exchange rates (24, 27). For example, AT base pairs in AT tracts have anomalously long lifetimes (≥ 100 ms) (28), and GC base pairs in GC tracts can show much shorter lifetimes than in other DNA sequences (27). These anomalies have been attributed to differential stacking interactions between various bases, steric conflicts within certain sequences, and differences in hydration of the DNA grooves (27).

In general, the rate constants and equilibrium constants for base pair opening are log-linearly correlated (Fig. 4), indicating that the strength of the base pair determines the rate constant for opening. This linear relationship extends to mismatched bases (see T:G mismatch in Fig. 4) (29), which suggests that damaged bases are kinetically and thermodynamically more likely to be found in an extrahelical state. The observation that mismatched T:G base pairs open spontaneously with rate constants greater than or equal to the known rates of enzymatic base flipping ($\sim 1,000$ s^{-1}; see following text) (30–33) suggests that enzymes need not accelerate the opening of such destabilized bases. In contrast, an enzyme would gain tremendous advantage by increasing the small equilibrium constant for base pair opening ($K_{open} = 10^{-3}$ for a T:G base pair; Fig. 4) (29). A change in the equilibrium constant could, in principle, occur by destabilizing the duplex, forming strong interactions with the extrahelical base, or a combination of both strategies (see following text).

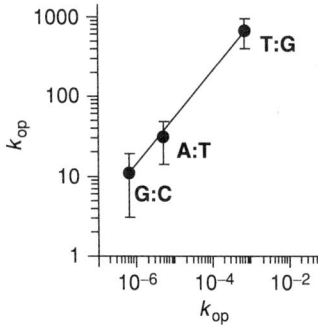

FIG. 4. The opening rates and equilibrium opening constants for DNA base pairs are log-linearly correlated. This relationship indicates that intrinsic base pair stability could affect both the kinetics and thermodynamics of enzymatic base flipping (see text).

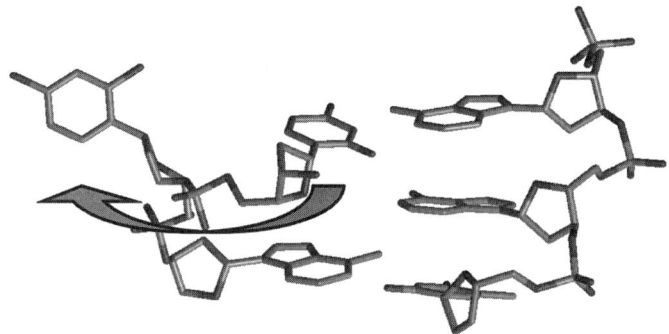

FIG. 5. Predicted pathway for spontaneous flipping of a uracil base. The final extrahelical structure was approximated from the potential-of-mean-force calculations of Banavali and MacKerell (26). These calculations indicate that the energy surface for base flipping is quite flat once the base pair hydrogen bonds are disrupted, and that the base may exit from the major or minor grooves, which is facilitated by localized distortions in the phosphodiester backbone around the flipped base.

B. Computational Studies

Computational approaches have been used to investigate the structural transformations that occur along the pathway for spontaneous and enzymatic base pair opening (26, 34). The most comprehensive study mapped the complete pathway for flipping a base from the major or minor groove of B DNA (26), from which it was concluded that both exodus routes have nearly identical activation barriers ($\sim$15 to 20 kcal/mol). The path of the migrating base was found to be nearly perpendicular to the helical axis (Fig. 5), tracking closely along the chosen exit groove. Only small, localized phosphodiester backbone dihedral distortions were required to allow base flipping, but the same distortions were observed for both the minor and major groove paths. The requirement for DNA backbone distortions suggests that enzymes might facilitate base flipping by using binding energy to induce similar alterations in the DNA backbone to assist in extricating the base from the duplex.

III. Enzymatic Base Flipping

The last eight years have seen an explosion of crystallographic structures of DNA glycosylases bound to their respective DNA substrates (Fig. 1B). These studies have revealed unifying features of these complexes that together suggest some critical structural transformations that must occur during the process of base flipping. These structural studies, which only describe the final, most stable flipped-out state, have been reinforced by spectroscopic studies identifying transient intermediates that ultimately lead to the final extrahelical state. Together, these approaches suggest a series of rapid, stepwise structural transformations that serve as essential specificity gates preventing productive flipping of normal bases.

A. Structural Studies

Although the first structure of an enzyme-DNA complex showing an extrahelical base was that of Hha I cytosine-5-methyltransferase (6), the first structure of a DNA glycosylase bound to a stable substrate analogue was that of UDG complexed with the C-glycoside, pseudodeoxyuridine (10). Subsequently, structures of many other DNA glycosylase complexes have been reported, and each shows a similar structural mode of base flipping as first reported for UDG. These enzymes include the alkyladenine DNA glycosylase II from *E. coli* and its functional homologue from humans (AAG) (13, 15, 35), the 8-oxoguanine DNA glycosylase from *E. coli* (MutM) and humans (hOGG1) (18–20), and the pyrimidine dimer DNA glycosylase from T4 phage (36). These structures have shown the following highly conserved features associated with flipping bases: DNA bending (Fig. 6A), distortion of the phosphodiester dihedral angles on the 3′ and 5′ sides of the extrahelical nucleotide (Fig. 6B), protrusion of a bulky residue into the minor groove of the DNA to fill the void once occupied by the expelled base (Fig. 6B), and, in some cases, the formation of damaged base contacts that are required for induced-fit specificity (Fig. 6C). In addition, minor differences in nonspecific enzyme–phosphodiester interactions on the flipped and unflipped DNA strands can be observed in these structures. A key step in understanding base flipping, and the specificity it provides, is to determine the temporal sequence for forming these interactions, and their energetic importance at each step along the flipping pathway. These goals have been partially achieved for UDG.

B. Methods and Mechanism

Perhaps the most important tool that has been utilized to probe the process of base flipping in several systems is the fluorescent adenine analogue, 2-aminopurine (2-AP; Fig. 7A) (30, 32, 33, 37, 38). 2-AP can be incorporated into DNA by solid phase synthesis and provides a strong spectroscopic signal to monitor the real-time kinetics and thermodynamics of base flipping. Typically, 2-AP is placed adjacent to the base that is flipped by the enzyme, and the process of flipping the damaged base gives rise to a large increase in 2-AP fluorescence at 370 nm (Fig. 7B). The physical basis for the large fluorescence increase of 2-AP upon base flipping has been extensively studied (39–43), and results in part from disruption of the effects of static and dynamic quenching when 2-AP is stacked with other bases in duplex DNA. A second useful

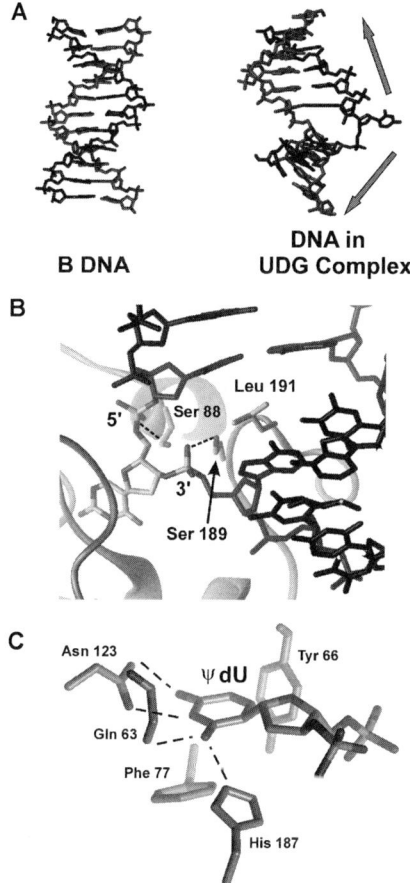

FIG. 6. General aspects of enzymatic base flipping as exemplified by the enzyme UDG. (A) All enzymes that flip bases are found to bend DNA in the range (22 to 75°) (56). Bending may be one of the earliest events on the base-flipping pathway that leads to destabilization of the base pair and also provides a sterically unobstructed exit route for the base. (B) The UDG–DNA interactions have been shown to be important for ejecting the uracil base into the enzyme active site. The Leu 191 wedge that pushes the base from the duplex and the two serine residues that pinch the phosphodiester backbone on either side of the flipped deoxyuridine nucleotide are shown. (C) Specific hydrogen bonding and aromatic stacking interactions with the pseudodeoxyuridine substrate analogue in the final flipped state (E°F, Figure 8). The residue numbering corresponds to the *E. coli* enzyme.

approach that has been recently used to study the mechanism of base flipping is atomic force microscopy (44). This method can provide quantitative measurements on the DNA bending transformations that accompany enzyme

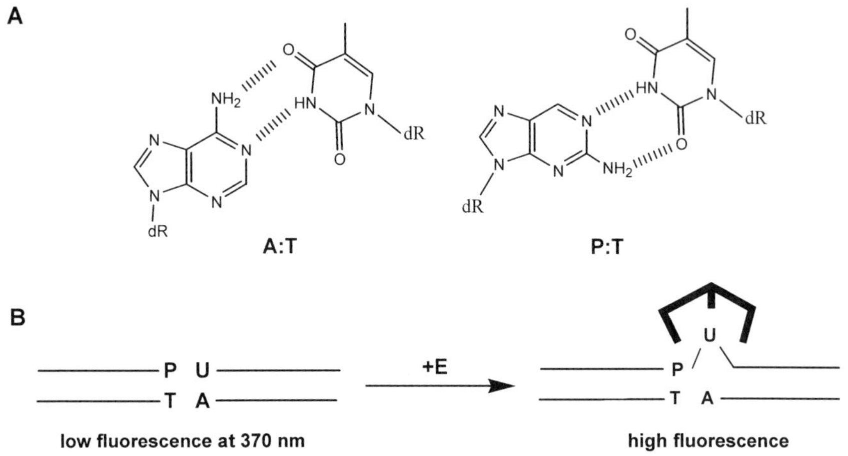

FIG. 7. The adenine analogue, 2-aminopurine (P), is a nonperturbing and useful fluorescent reporter group for studying base flipping in real time. (A) Like adenine, 2-aminopurine forms two hydrogen bonds with thymine. However, the 2-amino group is poised on the minor groove face of the helix, which differs from the 6-amino group of adenine. (B) 2-Aminopurine is a valuable probe for base flipping because its fluorescence is highly quenched when it is stacked in the duplex DNA environment, and increases by 10- to 100-fold when it is exposed to solvent (37, 42, 88). Accordingly, a large 8-fold increase in 2-aminopurine fluorescence is observed when the adjacent uracil is flipped into the UDG active site (30).

binding to both nonspecific and damaged sites in duplex DNA constructs. It may also provide the relative populations of the nonspecific and specific complexes, which may be related to the relative binding affinity of the enzyme to these sites in solution. Such approaches have indicated that base flipping is a multistep process that involves (i) nonspecific DNA binding with induced destabilization by bending or other mechanisms (Fig. 8, ES), (ii) flipping of the damaged base into a metastable flipped state that is not yet fully docked into the enzyme active site (Fig. 8, EF), and (iii) a rate-limiting conformational change in the enzyme that leads to formation of the specific stacking and hydrogen bonding interactions that promote stable binding of the extrahelical base (Fig. 8, E°F), and ultimately, glycosidic bond cleavage.

1. The Earliest Events: Search Mechanisms and Nonspecific DNA Destabilization

One of the most interesting questions about the mechanism of base flipping is how the whole process gets started. On the basis of the DNA bending observed in all the crystal structures of enzymes with flipped-out bases, it has been speculated that the earliest event may be bending of DNA when the enzyme forms the initial nonspecific complex (44, 45). This is an

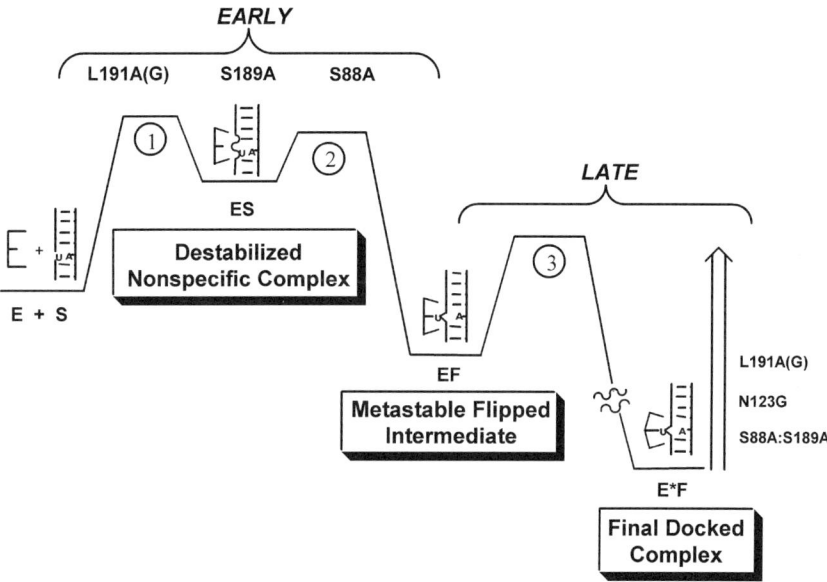

FIG. 8. Stepwise base flipping mechanism for UDG (30–33). Rapid kinetic and mutagenesis studies have uncovered three distinct steps in the base flipping pathway for UDG: (1) encounter of the enzyme with the DNA to form a destabilized nonspecific complex which occurs at a diffusion-controlled rate ($k_1 = 2.2 \times 10^8 \text{M}^{-1}\text{s}^{-1}$, $k_{-1} = 600\,\text{s}^{-1}$), (2) extrusion of the uracil base from the duplex to form a weakly bound and unstable extrahelical intermediate (EF, $k_2 = 700\,\text{s}^{-1}$, $k_{-2} = 180\,\text{s}^{-1}$), and (3) formation of the final fully docked complex which requires isomerization of UDG from an open to closed form ($\text{E}°\text{F}$, $k_3 = 350\,\text{s}^{-1}$, $k_{-3} = 100\,\text{s}^{-1}$). The effects of mutations (see text and Fig. 6) can be considered as early or late. Thus, mutations that affect the overall rate of formation, or stability, of the EF complex are termed early, and those that affect the E°F complex are termed late. The arrow depicts the strong destabilizing effects of the indicated mutations on the E°F complex. The barrier heights are proportional to the kinetic constants rather than the activation energies to better reveal the energetic profile of the pathway.

attractive proposal that is compatible with computational studies, suggesting that alterations in the DNA phosphate backbone dihedral angles (26), and/or widening of the DNA major groove (46), would facilitate base flipping. A second important facet of this destabilization mechanism is the relative energetics for expelling damaged and undamaged bases from the duplex, which might provide the first handle for discriminating between these bases. Since most DNA lesions and mismatches are destabilizing (see for example the T:G mismatch in Fig. 4), it is reasonable to expect that damaged sites would be easier to distort due to their weaker hydrogen bonding and stacking, or their increased flexibility (42, 43, 47, 48), all of which would lead to more facile flipping of damaged bases. In summary, the intrinsic deformability of damaged sites, coupled with the enzymatic forces that destabilize the duplex structure, are likely two important aspects of the earliest events in damaged base flipping.

One related issue that continues to be debated is whether such DNA distortions might also lead to flipping of normal DNA bases during a processive enzymatic scan for damaged bases in DNA (49). Verdine originally proposed that DNA glycosylases use a unique processive mechanism to find damaged bases, in which undamaged bases are reversibly flipped into the enzyme active site and then inspected for their integrity before the enzyme moves along the duplex to flip the next base (49). Although such a mechanism is intrinsically complicated, and suggests very complex motions of the enzyme and DNA during the search process, it was argued that, in thermodynamic terms, the energetic cost of flipping each base would be paid for upon its return to the DNA base stack during each flipping event. The major tenet of this proposal—that glycosylases can flip normal bases—is supported by the observation that some DNA glycosylases are capable of cleaving the glycosidic bond of normal bases (albeit very slowly), suggesting that normal bases can be flipped into the active sites of some enzymes (50). In contrast, fluorescence studies of UDG have not found any evidence for flipping of any base other than uracil, although mutant UDG enzymes have been constructed that are capable of cleaving cytosine and thymidine (51, 52). The observed activity of these mutant enzymes is large enough that they must certainly take advantage of the potent active site chemistry of UDG, and therefore, must also flip these bases. The wild-type UDG active site sterically excludes all other normal bases, and in addition, prevents productive binding of these bases through a series of conformational changes that can only be provoked by the uracil base (30, 53–55).

Although the processive flipping mechanism is plausible on the basis that some enzymes flip normal bases, one limitation of this mechanism is that it takes time to flip bases (see following text), and doing so repeatedly could actually increase the time required to locate the specific site as compared to alternative random diffusion site location mechanisms (30). Another requirement for the viability of this processive search mechanism is that the nonspecific complex should have a dissociation rate (k_{off}) that is slow compared to the translocation rate down the DNA (k_{trans}) (Fig. 9). If this requirement is not met, and the probability (P) of dissociating from the DNA is large [$P = k_{off}/(k_{trans} + k_{off}) \gg 1$], then little processivity will be possible. Alternatively, if $k_{off}/(k_{trans} + k_{off})$ approaches zero, then the enzyme is permanently trapped on the DNA, and a processive search mechanism must be followed.

Is there any experimental evidence to suggest whether processivity plays a significant role in glycosylase-DNA interactions? Experimental measurements for 8-oxoguanine DNA glycosylase (hOGG1) and UDG indicate that nonspecific binding is very weak ($K_D \sim 10$ to $40 \ \mu M$) (30, 33). In addition, an extremely rapid dissociation rate of about 1000 to 5000 s^{-1} has been estimated

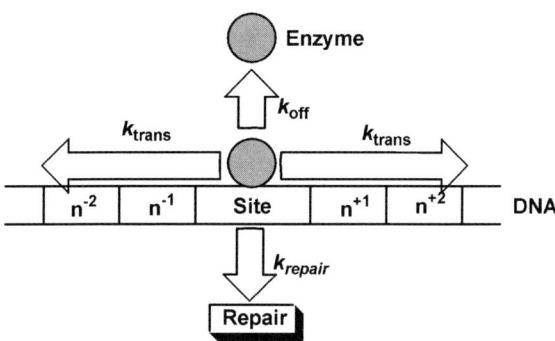

FIG. 9. Kinetic partitioning determines the processive or distributive nature of the damage search process. If the dissociation rate of the enzyme from the DNA (k_{off}) is much greater than the translocation rate along the DNA (k_{trans}), then the enzyme cannot use a processive search for the damaged site because the probability for dissociation is large [$P = k_{off}/(k_{trans} + k_{off}) \gg 1$]. Even if k_{trans} is large enough to allow a processive search, such a rapid rate would be a hindrance to repair because the rate constant for repair (k_{repair}) would have to be of comparable magnitude to make the probability of repair high [$P = k_{repair}/(k_{trans} + k_{repair} + k_{off})$]. For UDG, the exceptionally large off rate from nonspecific DNA makes a processive mechanism physically unreasonable [see text and (56)].

for the nonspecific UDG–DNA complex (30, 33). Neither of these results is consistent with a highly processive search mechanism unless k_{trans} is $\gg$100,000 s^{-1}. As previously noted (56), a high sliding rate (if occurring) would be counterproductive once the damaged site is located because the probability of repair (P) depends on the rate constant for repair of the site (k_{repair}), k_{trans} and k_{off} (i.e., $P = k_{repair}/(k_{trans} + k_{repair} + k_{off})$. Thus for large sliding rates, k_{repair} is much less than k_{trans}, and the efficiency of repair is exceedingly low. Although numerous studies have been performed to address the processivity of DNA repair glycosylases, none of these has revealed a large amount of processivity at physiological salt concentrations (57–61). New improved approaches for investigating site location mechanisms have not yet been applied to DNA glycosylases (62, 63), but should eventually uncover the nature of the search strategy used by these enzymes.

What is the experimental evidence supporting duplex bending or destabilization upon nonspecific DNA binding? There are two systems that have been studied in sufficient detail to provide evidence for such mechanisms. The first is hOGG1, where atomic force microscopy has been used to examine whether this enzyme bends DNA when it binds to nonspecific sites (44). Using linearized plasmid DNAs containing either a single 8-oxoguanine base or no damaged base at all, Verdine and coworkers collected AFM images of hOGG1–DNA complexes. The images clearly confirmed the crystallographic observation that hOGG1 bends DNA when it is bound to an 8-oxoguanine site, but more importantly, also showed that hOGG1 induces similar bending when it binds to nonspecific sites ($\sim 72°$). From the relative populations of bent and unbent nonspecific DNA complexes, a bending equilibrium (K_{bend}) of about 2 was calculated (K_{bend} = [bent DNA complexes]/[unbent DNA complexes]). The authors suggested that the observed DNA bending would not be structurally feasible without nonspecific base flipping, and concluded that hOGG flips all bases, as required in the processive search mechanism. Although bending is clearly indicated in this study, the AFM results cannot discriminate whether a processive or distributive search mechanism for the damaged site is followed (see previous text).

Kinetic and thermodynamic studies of UDG have also provided evidence for transient DNA destabilization upon nonspecific binding of the enzyme. As presented above and in Fig. 8, uracil flipping is a multistep process in which the local structure of the DNA around the flipped base is progressively changed from B DNA to the final flipped structure. From the observation that the internal equilibrium for forming the flipped-state in the UDG reaction was independent of whether the uracil was originally located in a stable U/A base pair, a less stable wobble U/G base pair, or even ssDNA, it was concluded that the intrinsic energetic differences between these DNAs with respect to extracting the uracil must have been removed upon formation of the very first nonspecific complex (ES, Fig. 8). Thus, UDG apparently uses binding energy to destabilize the duplex before beginning the process of flipping the base. Unlike the AFM studies with hOGG1, solution NMR structural studies of nonspecific DNA bound to UDG show no large changes from typical B DNA structure (Cao and Stivers, unpublished). Thus, the DNA destabilization induced by UDG is likely to cause subtle changes in the DNA structure, or the destabilized state is of high energy and not significantly populated in solution. These NMR results predict that bent DNA would not be observed in AFM studies of UDG complexed with nonspecific DNA.

2. The Metastable Flipped State

After locating the site and forming the weak nonspecific complex in the correct register to flip the damaged base, UDG then expels the uracil into a partially flipped state. This state is poorly characterized in terms of its structure, but has been detected in stopped-flow fluorescence kinetic studies in which the 2-AP fluorescence of the DNA and the tryptophan fluorescence of the enzyme were both monitored (Fig. 8). Both fluorescence signals showed saturation kinetics, indicating that there was a change in rate-limiting step from association of the enzyme with the DNA, to a second internal isomerization step, the rate of which was independent of reactant concentration (31, 32). However, the isomerization step that was detected by tryptophan fluorescence was about two times slower than the isomerization rate detected by 2-AP fluorescence. Since 2-AP fluorescence probes the disruption of base stacking interactions, and tryptophan fluorescence probes the conformation of the enzyme, then the data strongly suggested that rapid base pair disruption preceded a slower conformational change in UDG. The relative populations of the three enzyme-DNA complexes were estimated from simulations of the kinetic and thermodynamic data: nonspecific complex (ES = 5%), metastable flipped complex (EF = 21%), and fully flipped complex (E°F = 74%) (see Fig. 8). Thus, UDG follows a stepwise pathway to form the final state, which provides two gating barriers to prevent the adventitious cleavage of undamaged bases.

Extensive mutagenesis and kinetic studies have uncovered three enzyme residues of UDG that are important for forming the metastable flipped state. The first residue is Leu 191, which is observed to protrude into the minor groove of the DNA substrate, and occupy the space originally filled by the uracil base (31, 32, 64) (Fig. 6). Upon binding to a damaged site, wtUDG undergoes a conformational closing that gives rise to a significant movement of Leu 191, pushing its side chain into the minor groove where it acts as a molecular wedge to pry the uracil residue from the base stack. Accordingly, mutation of Leu 191 to alanine or glycine results in an 8- to 16-fold slower association rate of UDG with damaged DNA, and most importantly, also severely destabilizes the fully flipped state. These two distinct effects of removing Leu 191 indicate that this residue plays a role in helping UDG rapidly associate with damaged DNA (the "wedge" function), and then stabilize the base once it is fully inserted in the active site pocket (a "plugging" function) (31, 32, 64, 65).

Removal of the two serine residues that interact with the phosphodiester groups on both sides of the flipped deoxyuridine also causes an order of magnitude decrease in the association rate of UDG with damaged DNA (Fig. 6) (32). These groups may distort the DNA backbone at the very earliest steps of the flipping process, thereby destabilizing the uracil in its duplex environment and widening the exit groove (10, 45). As flipping proceeds, these side chains likely further compress the DNA backbone, giving rise to the highly unusual phosphodiester dihedral angles observed in the crystal structures of UDG (Fig. 6). Both serine residues are required for stabilization of the final flipped state, as their removal totally ablates formation of this state (32). Thus, nonspecific serine–phosphodiester interactions are critical both in the early and latest stages of base flipping.

3. THE FINAL FULLY DOCKED STATE

As has been noted, the final kinetic gate leading to the fully flipped state involves a conformation change in UDG, and requires the Leu 191 wedge as well as both serine side chain interactions. Once the base is docked deep in the active site, the final specific hydrogen bonds between uracil 3-H, O2, and O4 and enzyme side chain and backbone atoms form. Two of these enzyme side chains have been changed to alanine or glycine (Asn 123 and His 187), and the effects on base flipping were assessed (*32*). In contrast to the L191A, S88A, and S189A mutations, removal of the Asn 123 and His 187 had little effect on the early step of forming the metastable flipped intermediate, indicating that these interactions form very late in the flipping process, most likely when the final docking takes place. The N123G mutation severely destabilized the final state, and showed no evidence for the induced fit conformational change that gives rise to this state. In contrast, the H187A mutant bound to DNA very similarly as wtUDG, and achieved the final state with similar kinetics. This result is consistent with previous steady-state and single turnover kinetic studies of the H187A mutant, which established that this residue does not form strong interactions with the uracil base until the transition-state is reached (*54, 66, 67*). The picture that emerges from these studies is one where strong energetic coupling exists between residues that form the early interactions (Leu 191, Ser 88, and Ser 189), and those that exclusively form specific contacts very late in the flipping process (Asn 123 and His 187). This mechanism provides specificity because the early contacts are required to form the late specific contacts, and the late contacts can only be fulfilled by the uracil base.

4. Atypical Base Flippers

Although the vast majority of enzymes that flip bases show similar structural transformations as UDG and hOGG1, there are two enzymes that are quite distinct. The first is the pyrimidine dimer DNA glycosylase (PDG) from T4 phage, which acts to remove cis-syn cyclobutane pyrimidine dimers that result from ultraviolet irradiation of DNA (*68*). Strikingly, PDG is found to flip not the pyrimidine dimer itself, but the adenine that is located opposite to the 5′ thymine of the dimer (Fig. 10) (*36*). Although flipping by PDG also involves DNA bending and contacts between the enzyme and several phosphodiester groups around the lesion, T4 PDG does not use a wedge group to push the adenine from the base stack, which is a striking deviation from other enzymes. Instead of being thrust into a highly specific pocket that only accommodates adenine, the extrahelical base is held in a featureless protein crevice by what must be assumed to be weak van der Waals forces. Overall, T4 PDG appears to have evolved a different solution to gain essential access to the glycosidic linkage, because the rigidity of the pyrimidine dimer makes direct lesion flipping an unworkable alternative.

Since nature has provided examples of direct lesion flipping, as well as flipping of the opposing base, it perhaps should not be surprising that an example of "double flipping" exists: the MutY adenine DNA glycosylase (*69, 70*). MutY

DNA DAMAGE RECOGNITION AND BASE FLIPPING

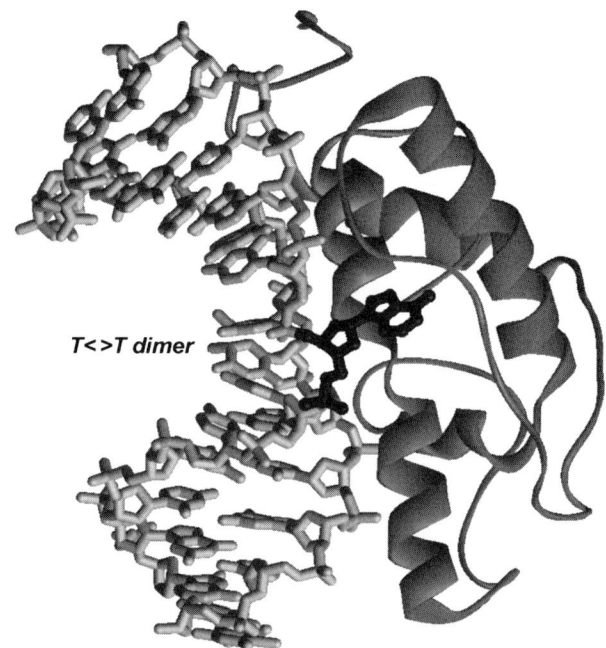

FIG. 10. Structure of the pyrimidine dimer DNA glycosylase from phage T4 (36). This enzyme is found to flip the adenine base opposite to the T<>T pyrimidine dimer, as an alternate strategy for gaining access to the damaged site.

excises the normal base adenine when it is has become misincorporated by DNA polymerase opposite to the oxidized base 8-oxoguanine (71). The enzyme consists of two domains, a 26 kDa catalytic core domain that contains a very specific binding pocket for an extrahelical adenine (72), and a 13 kDa carboxyl terminal domain that binds 8-oxoguanine (73, 74). A complete structure of MutY has not yet been reported, but a crystal structure of the catalytic domain (72) and a homology model of the smaller 8-oxoguanine binding domain have been obtained (73, 74). The enzyme apparently wraps around the DNA duplex allowing the carboxyl domain to rapidly flip the 8-oxoguanine base, followed much more slowly by adenine flipping into the active site of the catalytic domain (69). As observed with UDG, a slower isomerization of the enzyme completes the flipping process, leading to exceptionally tight binding of the double-flipped DNA (69). This remarkable mechanism, involving flipping of the opposing 8-oxoguanine base, appears to provide an ironclad insurance policy against the inadvertent removal of adenine when it is paired with its usual partner thymidine.

IV. New Experimental Approaches

A. Preorganized Substrates

The act of flipping a base from the DNA duplex would be expected to involve a significant unfavorable energetic penalty due to the disruption of favorable hydrogen bonding and stacking interactions of the base when it is located in the duplex environment. Layered on top of these localized changes in DNA structure is the energetic cost of DNA bending that must also be paid for with favorable binding interactions of the enzyme with the DNA. These energetic considerations led to the general concept that preorganizing a base in a flipped-out conformation would lead to tighter binding to the specific site. The simplest mechanism by which increased affinity could be brought about is increasing the effective concentration of the reactive extrahelical conformation, which would lead to an increased association rate constant of the enzyme with the specific site.

So what tools are available to bring about preorganization of base in an extrahelical state? For UDG, one recent approach was to incorporate a bulky pyrene nucleotide (Y) opposite to the uracil base in duplex DNA (Fig. 11) (31, 64). The unnatural pyrene nucleotide fills the space normally taken up by an entire base pair, leading to partial or complete expulsion of the uracil base from the DNA base stack, which was established by its high sensitivity to

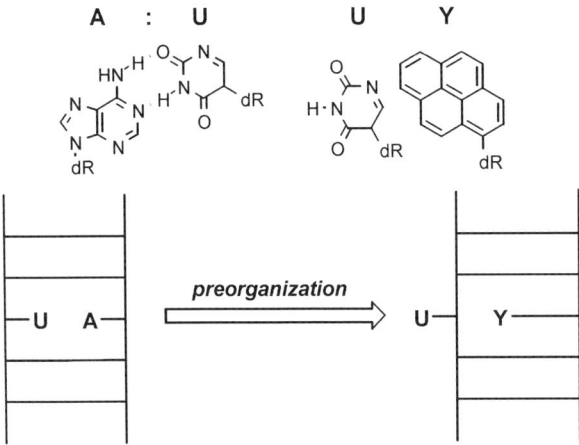

FIG. 11. Enhanced enzyme recognition using a substrate with a preorganized extrahelical base. Incorporation of an unnatural pyrene nucleotide opposite to uracil in duplex DNA leads to expulsion of the uracil due to the large surface area of pyrene that fills the entire DNA base stack normally occupied by uracil and adenine (see text).

DNA DAMAGE RECOGNITION AND BASE FLIPPING 55

oxidation by potassium permanganate. Rapid kinetic and binding studies with wtUDG confirmed that the enzyme bound to the U/Y duplex with 10-fold greater affinity than that of the normal duplex containing a U/A base pair, and that most of this increase was brought about by increasing the association rate 5-fold to 5×10^8 M^{-1} s^{-1}. This rate constant is at the diffusion-controlled limit and suggests that every encounter of the enzyme leads to productive docking of the uracil base into binding pocket. As might be expected from preorganization, the U/Y analogue did not show any evidence for formation of the metastable flipped state, and instead, rapidly formed the final fully docked state. These findings led to the conclusion that pyrene had changed the recognition mechanism from induced-fit to lock-and-key.

A further aspect of the pyrene nucleotide "wedge" is its ability to rescue UDG mutations that prevent base flipping (31, 32). One of the most striking examples was the complete rescue of the damaging effects of the L191A and L191G mutations when the pyrene substrate was utilized. As has been described, these mutants are completely incapable of attaining the extra-helical state with the U/A substrate (E°F, Fig. 8), but are kinetically and thermodynamically indistinguishable from wtUDG when the U/Y wedge substrate is employed. In other words, the removal of the Leu 191 wedge side chain has made these enzymes specific for U/Y base pairs. An extension of this approach led to the construction of a new UDG double mutant (N123D:L191A, Fig. 12) that recognizes cytosine when it is paired with pyrene, yet does not remove cytosine from a normal C/G base pair (75). The successful design of this mutant suggests that similar chemical strategies may used to target other flipping enzymes to specific sites in DNA. We have described new improved syntheses of aromatic C-nucleosides that allow such molecules to be obtained in high yield and in the correct anomeric configuration (76, 77).

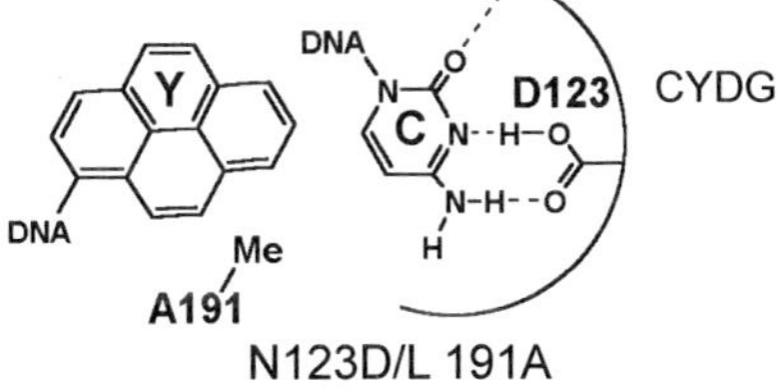

FIG. 12. Rational engineering of the UDG active site to produce a new glycosylase with specificity for cytosine:pyrene base pairs. Two mutations were introduced into UDG: the N123D mutation creates a new residue that can hydrogen bond with the Watson-Crick donor acceptor group of cytosine, while the L191A mutation makes the glycosylase activity dependent on pyrene assistance (see text). Thus, a new enzyme (cytosine-pyrene DNA glycosylase, CYDG) was created.

B. Nonpolar Damaged Base Isosteres

One key question in understanding the factors that determine whether a base is flipped by an enzyme is the role of the specific hydrogen bonds that are broken in the DNA substrate and then reformed upon docking of the base in the enzyme active site. The effective chemical tools for addressing such a question were first developed by Kool for investigation of hydrogen bonding and base shape in recognition by DNA polymerases (78–80). We have used a similar approach with UDG by substituting a 2, 4 difluorophenyl nucleotide (F) for deoxyuridine in duplex DNA (Fig. 13). The fluorine atoms of F are isosteric with the oxygen atoms of uracil, yet are incapable of forming anything but the weakest hydrogen bonding interactions with donor groups in the DNA or enzyme active site (81). These attributes of F allow the investigation of whether UDG can flip a base from DNA when its exit is not impeded by the energetic cost of hydrogen bond disruption. In addition, the ^{19}F atoms provide valuable spectroscopic signals for NMR spectroscopic studies of the base flipping process.

Binding studies of identical duplex DNAs containing either an F/A or U/A base pair revealed that F/A binds only 5-fold more weakly than the U/A analogue (Jiang and Stivers, submitted), indicating that the net energetic effect of removing the uracil hydrogen bonding groups, while retaining the stacking interactions, is quite small. This result suggests that the energetic cost of breaking the hydrogen bonds in the U/A base pair is nearly compensated for by the new hydrogen bonds that are formed in the UDG active site. This conclusion assumes that F is flipped out of the duplex and docked into the UDG active site in a similar fashion as uracil. Evidence that this was the case was obtained from 1D ^{19}F NMR studies of the F/A duplex free in solution and bound to UDG (Fig. 14). The fluorine shifts of F show a diagnostic upfield shift when stacked in a duplex environment, and move as much as 2 ppm

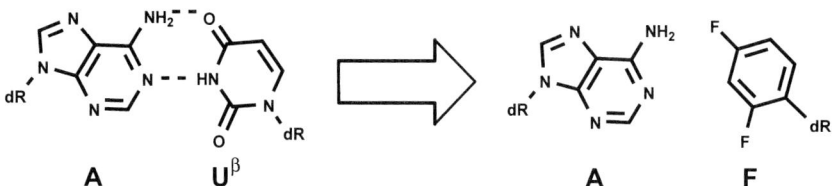

FIG. 13. Utilization of a nonpolar uracil isostere, 2, 4, difluorophenyl nucleotide (F) to probe the requirement for specific hydrogen bonding interactions in base flipping. The fluorine atoms of F are isosteric with the exocyclid oxygen atoms of uracil, yet are not capable of forming hydrogen bonds in the DNA base stack or in the UDG active site. Binding and ^{19}F NMR studies (Fig. 14) have shown that F is flipped from the duplex by UDG and binds only 5-fold less tightly than U^{β} substrate analogue DNA.

DNA DAMAGE RECOGNITION AND BASE FLIPPING 57

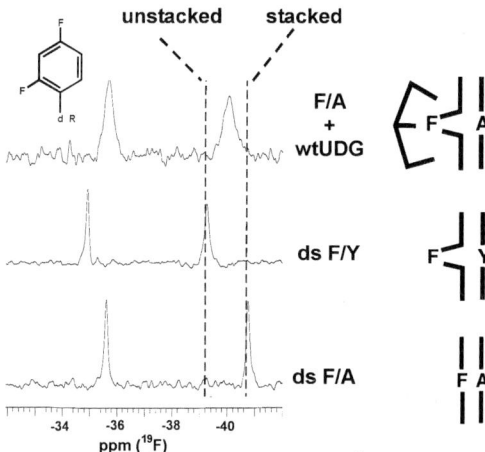

FIG. 14. ^{19}F NMR studies of 2, 4 difluorophenyl nucleotide (F) flipping by UDG. When stacked in duplex DNA opposite to an adenine base (ds F/A), the two fluorine atoms of F resonate at higher field than when forced into an extrahelical state in an F/Y base pair (ds F/Y, middle spectrum) or in the context of single-stranded DNA (ss F, not shown). When UDG binds to the ds F/A duplex (top spectrum), the fluorine atoms are shifted downfield chemical shifts that are intermediate between that observed in the ds F/A duplex and the ds F/Y DNA. This is consistent with F attaining the metastable state in which the base is in a dynamic equilibrium between a flipped state and the intrahelical state. As would be expected, ds F/A DNA does not induce a tryptophan fluorescence change in UDG that is diagnostic for attaining the E°F state (see Fig. 8). These inferences from chemical shift changes were directly confirmed in REDOR NMR experiments in which dipolar interactions between the fluorine atoms of F and several side chain ^{15}N nuclei of UDG were measured (McDowell, L., Schaeffer, J., and Stivers, J. T., unpublished).

downfield when the base becomes unstacked upon UDG binding. This mechanistic conclusion that a downfield fluorine shift reflects unstacking is supported by the similar downfield migration of the fluorine shift when F is located in single-stranded DNA or in duplex DNA containing an F/Y base pair. Further support that extrahelical interactions exist between UDG and F was provided by the observation that removal of the L191A wedge weakened binding to the F/A duplex by the same amount as a U/A duplex, and the mutational effect could be rescued by pyrene in the same fashion as has been described for the U/Y substrate (i.e., an F/Y base pair). Interestingly, it was found that binding of duplex DNA containing an F/A base pair does not lead to UDG tryptophan quenching, suggesting that F never achieves the fully docked E°F state (Fig. 8). These results indicate that F is arrested at an extrahelical state that mimics the metastable state detected in stopped-flow experiments with uracil-containing DNA (EF, Fig. 8). This is not surprising because the mutagenesis experiments indicated that hydrogen bonding interactions with the base are required to stabilize the final flipped-out state (E°F, Fig. 8), and F cannot provide these interactions. These findings with the isosteric F nucleotide show that steric complementarity between the flipped base and the active site is sufficient to allow partial but not full flipping.

C. Damaged Base Pair Stability and Specific Recognition

A question of continuing interest is what features of the damaged site facilitate enzyme binding and base flipping, and whether the initial step involves extrahelical or intrahelical recognition of the damaged base. If extrahelical recognition is required to detect the damage, then this would imply that repair enzymes must in some way at least partially expel normal bases to distinguish these bases from the desired target base. Proposals for specific recognition invoke destabilization brought about by weaker hydrogen bonding of some damaged base pairs as compared to normal bases, the disruption of DNA base stacking by the damage, and the possible increased flexibility of damaged sites in DNA (43, 47, 48, 82). We have been investigating some of these proposals with UDG by making systematic changes in the base pair partner of the cognate uracil base, thereby incrementally destabilizing the uracil base in the DNA duplex (Fig. 15). A goal of this work is to quantitatively assess whether localized destabilization of the duplex correlates with increased affinity of the enzyme for the uracil base. Thus, these experiments look at one side of the specificity issue by addressing how much specificity can be provided by the intrinsic stability of the damaged site.

To experimentally address this question, a series of DNA substrate analogues containing U^β/X base pairs was constructed as depicted in Fig. 15. The hydrogen bonding interactions of these constructs have been incrementally removed from the base pair, allowing the energetic effect on duplex stability to be measured by differential scanning calorimetry, and then compared with the energetic effect on UDG binding. A plot of $\Delta\Delta H$ for duplex melting against $\Delta\Delta G$ for UDG binding was linear over a 3.5 kcal change in binding energy ($m = 0.06$, $r = 0.94$), providing solid evidence that destabilizing the duplex by removing hydrogen bonding interactions facilitates binding. In other words, less work has to be performed by UDG to flip U^β from a destabilized base pair, and some of this energy can be realized as tighter binding and specificity (89).

Additional experiments demonstrated that enhanced UDG binding also correlated with the solvent accessibility of the base. This conclusion was reached by investigating the sensitivity of thymidine in T:X base pairs to oxidation by permanganate (where X is one of the analogues shown in Fig. 15), and then comparing this parameter with the binding affinity of UDG for U^β:X containing DNA. (Thymidine rather than U^β was used in the permanganate sensitivity studies because the electron withdrawing 2′ fluorine

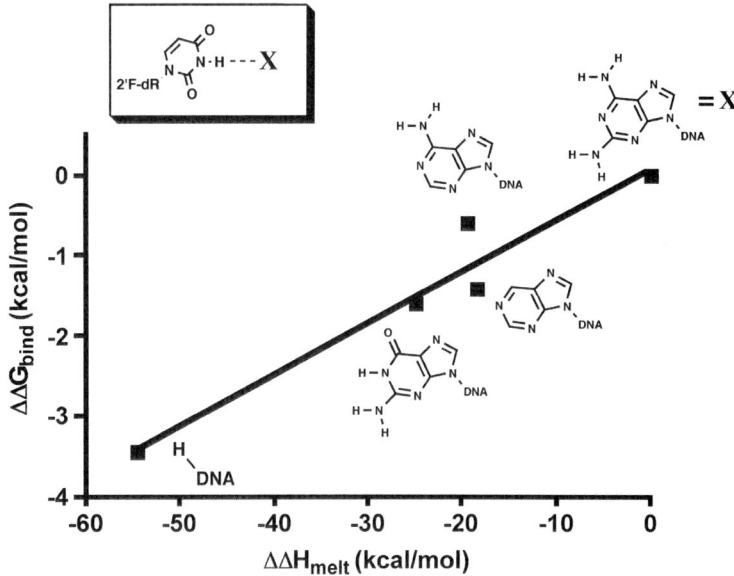

FIG. 15. Probing the role of base pair stability on uracil flipping by UDG (Krosky and Stivers, unpublished). Duplex DNA constructs were synthesized containing a stable 2′-fluorodeoxyuridine uracil base analogue (U^β) in one strand which was paired with a complementary strand containing a series of base analogues (X) that could form three, two, one, or no hydrogen bonds with U^β. The enthalpy of duplex melting was measured by calorimetry, and the free energy for binding of each DNA construct to UDG was also determined. The correlation shows that destabilizing the target base pair enhances target site binding by diminishing the energetic cost of base flipping, allowing UDG to use more of the available binding energy for tight binding.

atom of U^β interferes with the assay.) As observed with the enthalpy for DNA melting, there was a linear correlation between log S (where S is the fraction of the thymidine base in the T:X base pair that was oxidized) against log K_D for UDG binding (89). The correlation extended over 1.2 units in log S and 2.5 units in log K_D, with a slope of 0.4, and a correlation coefficient of 0.96. This correlation suggests that base pair breathing, which increases the concentration of the extrahelical base, may play a role in facilitating the rate of enzyme association with the site. This is similar to the effect of pyrene, which, in part, enhances the K_D for binding by increasing the concentration of the reactive extrahelical conformation. Despite these compelling correlations, it is important to point out that the maximum specificity contribution from damaged base pair destabilization (3.8 kcal/mol, Fig. 15) contributes only a small fraction of the total catalytic specificity of UDG ($\sim$10 kcal/mol) (21), the remainder being found in catalytic interactions in the transition-state.

V. Future Directions

The stage of the base flipping process that remains the most enigmatic and difficult to study experimentally is the initial nonspecific complex, which likely contains DNA that is perturbed in some way compared to the standard duplex form. Characterization of the structure and dynamics of the nonspecific complex would provide key insights into how the enzyme prepares the substrate for base flipping. It is doubtful that crystal structures will be obtained of these weak complexes without resorting to perturbations such as covalent cross-linking of the DNA and enzyme (83). If such structures are forthcoming, it will always be essential to establish that any proposed intermediates are on the pathway to achieving the final extrahelical state, using rigorous criteria such as kinetic competence as well as alternative solution experimental methods. In this respect, NMR spectroscopy will likely play a significant role in elucidating the structural features of such weak complexes. In addition, NMR can be used to study the dynamic behavior of the bound duplex using imino proton exchange methods (84) or other modern relaxation techniques that can reveal dynamic behavior over a wide range of time scales (85). On another front, the success with rational engineering of UDG to recognize specific cytosine residues in DNA using the pyrene substrate rescue method provides promise that other engineering projects will be successful, perhaps leading to new tools for molecular biology research or other applications. It is satisfying that fundamental research into an interesting problem in enzyme-DNA recognition can lead to basic principles that allow rational and successful manipulation of these systems. A recent crystal structure of MutY bound to damaged DNA does not provide evidence for a double-flip mechanism because only the adenine base of the 8-oxoG:A base pair was observed to be extrahelical (Fromme JC, Banerjee A, Huang SJ, Verdine GL. (2004) Structural basis for removal of adenine mispaired with 8-oxoguanine by MutY adenine DNA glycosylase. Nature **427**, 652–656). However, this does not negate the possibility that flipping of 8-oxoG occurs transiently on the flipping pathway, as suggested by biochemical experiments (69).

Acknowledgments

This work was supported by the National Institutes of Health grant GM56834 (to J. T. S.). I thank my collaborators over the years who have helped advance our knowledge on the mechanistic biology of DNA repair: Paul Carey at Case Western Reserve University, Gary Gilliland at the Center for Advanced Research in Biotechnology, Kris Pankiewicz at Pharmasset, Inc., and Yoshi Ichikawa at Optimer Pharmaceuticals. I also thank the members of my laboratory for their careful reading of this manuscript.

References

1. Seeberg, E., Eide, L., and Bjorås, M. (1995). The base excision repair pathway. *Trends Biochem. Sci.* **20**, 391–397.
2. Lindahl, T., and Andersson, A. (1972). Rate of chain breakage at apurinic sites in double-stranded deoxyribonucleic acid. *Biochemistry* **11**, 3618–3623.
3. Lindahl, T. (1990). Repair of intrinsic DNA lesions. *Mutat. Res.* **238**, 305–311.
4. Cuniasse, P., Fazakerley, G. V., Guschlbauer, W., Kaplan, B. E., and Sowers, L. C. (1990). The abasic site as a challenge to DNA polymerase. A nuclear magnetic resonance study of G, C, and T opposite a model abasic site. *J. Mol. Biol.* **213**, 303–314.
5. Slupphaug, G., Mol, C. D., Kavli, B., Arvai, A. S., Krokan, H. E., and Tainer, J. A. (1996). A nucleotide-flipping mechanism from the structure of human uracil-DNA glycosylase bound to DNA [see comments]. *Nature* **384**, 87–92.
6. Klimasauskas, S., Kumar, S., Roberts, R. J., and Cheng, X. (1994). HhaI methyltransferase flips its target base out of the DNA helix. *Cell* **76**, 357–369.
7. Roberts, R. J. (1995). On base flipping. *Cell* **82**, 9–12.
8. Lindahl, T., and Wood, R. D. (1999). Quality control by DNA repair. *Science* **286**, 1897–1905.
9. Lindahl, T., and Nyberg, B. (1974). Heat-induced deamination of cytosine residues in deoxyribonucleic acid. *Biochemistry* **13**, 3405–3410.
10. Parikh, S. S., Walcher, G., Jones, G. D., Slupphaug, G., Krokan, H. E., Blackburn, G. M., and Tainer, J. A. (2000). Uracil-DNA glycosylase-DNA substrate and product structures: Conformational strain promotes catalytic efficiency by coupled stereoelectronic effects. *Proc. Natl. Acad. Sci. USA* **97**, 5083–5088.
11. Daniels, D. S., Mol, C. D., Arvai, A. S., Kanugula, S., Pegg, A. E., and Tainer, J. A. (2000). Active and alkylated human AGT structures: A novel zinc site, inhibitor, and extrahelical base binding. *Embo. J.* **19**, 1719–1730.
12. Labahn, J., Scharer, O. D., Long, A., Ezaz-Nikpay, K., Verdine, G. L., and Ellenberger, T. E. (1996). Structural basis for the excision repair of alkylation-damaged DNA. *Cell* **86**, 321–329.
13. Lau, A. Y., Scharer, O. D., Samson, L., Verdine, G. L., and Ellenberger, T. (1998). Crystal structure of a human alkylbase-DNA repair enzyme complexed to DNA· Mechanisms for nucleotide flipping and base excision. *Cell* **95**, 249–258.
14. Lau, A. Y., Wyatt, M. D., Glassner, B. J., Samson, L. D., and Ellenberger, T. (2000). Molecular basis for discriminating between normal and damaged bases by the human alkyladenine glycosylase, AAG. *Proc. Natl. Acad. Sci. USA* **97**, 13573–13578.
15. Hollis, T., Ichikawa, Y., and Ellenberger, T. (2000). DNA bending and a flip-out mechanism for base excision by the helix–hairpin–helix DNA glycosylase, *Escherichia coli* AlkA. *EMBO J.* **19**, 758–766.
16. Hollis, T., Lau, A., and Ellenberger, T. (2000). Structural studies of human alkyladenine glycosylase and *E. coli* 3-methyladenine glycosylase. *Mutat. Res.* **460**, 201–210.
17. Nilsen, H., Rosewell, I., Robins, P., Skjelbred, C. F., Andersen, S., Slupphaug, G., Daly, G., Krokan, H. E., Lindahl, T., and Barnes, D. E. (2000). Uracil-DNA glycosylase (UNG)-deficient mice reveal a primary role of the enzyme during DNA replication. *Mol. Cell.* **5**, 1059–1065.
18. Bruner, S. D., Norman, D. P., and Verdine, G. L. (2000). Structural basis for recognition and repair of the endogenous mutagen 8-oxoguanine in DNA. *Nature* **403**, 859–866.
19. Fromme, J. C., and Verdine, G. L. (2002). Structural insights into lesion recognition and repair by the bacterial 8-oxoguanine DNA glycosylase MutM. *Nat. Struct. Biol.* **9**, 544–552.
20. Gilboa, R., Zharkov, D. O., Golan, G., Fernandes, A. S., Gerchman, S. E., Matz, E., Kycia, J. H., Grollman, A. P., and Shoham, G. (2002). Structure of formamidopyrimidine-DNA glycosylase covalently complexed to DNA. *J. Biol. Chem.* **277**, 19811–19816.

21. Stivers, J. T., and Jiang, Y. L. (2003). A mechanistic perspective on the chemistry of DNA repair glycosylases. *Chem. Rev.* **103**, 2729–2759.
22. Shapiro, R., and Danzig, M. (1972). Acidic hydrolysis of deoxycytidine and deoxyuridine derivatives. The general mechanism of deoxyribonucleoside hydrolysis. *Biochemistry* **11**, 23–29.
23. Zoltewicz, J. A., Clark, F. D., Sharpless, T. W., and Grahe, G. (1970). Kinetics and mechanisms of the acid-catalyzed hydrolysis of some purine nucleosides. *J. Am. Chem. Soc.* **92**, 1741–1750.
24. Gueron, M., Kochoyan, M., and Leroy, J. L. (1987). A single mode of DNA base-pair opening drives imino proton exchange. *Nature* **328**, 89–92.
25. Gueron, M., and Leroy, J.-L. (1995). Studies of base pair kinetics by NMR measurement of proton exchange. *Methods Enzymol.* **261**, 383–413.
26. Banavali, N. K., and MacKerell, A. D., Jr. (2002). Free energy and structural pathways of base flipping in a DNA GCGC containing sequence. *J. Mol. Biol.* **319**, 141–160.
27. Dornberger, U., Leijon, M., and Fritzsche, H. (1999). High base pair opening rates in tracts of GC base pairs. *J. Biol. Chem.* **274**, 6957–6962.
28. Moe, J. G., and Russu, I. M. (1990). Proton exchange and base pair opening kinetics in 5′-d(CGCGAATTCGCG)-3′ and related dodecamers. *Nucleic Acids Res.* **18**, 821–827.
29. Moe, J. G., and Russu, I. M. (1992). Kinetics and energetics of base-pair opening in 5′-d(CGCGAATTCGCG)-3′ and a substituted dodecamer containing G.T mismatches. *Biochemistry* **31**, 8421–8428.
30. Stivers, J. T., Pankiewicz, K. W., and Watanabe, K. A. (1999). Kinetic mechanism of damage site recognition and uracil flipping by *Escherichia coli* uracil DNA glycosylase. *Biochemistry* **38**, 952–963.
31. Jiang, Y. L., Song, F., and Stivers, J. T. (2002). Base flipping mutations of uracil DNA glycosylase: Substrate rescue using a pyrene nucleotide wedge. *Biochemistry* **41**, 11248–11254.
32. Jiang, Y. L., and Stivers, J. T. (2002). Mutational analysis of the base flipping mechanism of uracil DNA glycosylase. *Biochemistry* **41**, 11236–11247.
33. Wong, I., Lundquist, A. J., Bernards, A. S., and Mosbaugh, D. W. (2002). Presteady-state analysis of a single catalytic turnover by *Escherichia coli* uracil-DNA glycosylase reveals a "pinch–pull–push" mechanism. *J. Biol. Chem.* **20**, 20.
34. Huang, N., Banavali, N. K., and MacKerell, A. D., Jr. (2003). Protein-facilitated base flipping in DNA by cytosine-5-methyltransferase. *Proc. Natl. Acad. Sci. USA* **100**, 68–73.
35. Hollis, T., Lau, A., and Ellenberger, T. (2000). Structural studies of human alkyladenine glycosylase and *E. coli* 3-methyladenine glycosylase. *Mutat. Res.* **460**, 201–210.
36. Vassylyev, D. G., Kashiwagi, T., Mikami, Y., Ariyoshi, M., Iwai, S., Ohtsuka, E., and Morikawa, K. (1995). Atomic model of a pyrimidine dimer excision repair enzyme complexed with a DNA substrate: Structural basis for damaged DNA recognition. *Cell* **83**, 773–782.
37. Stivers, J. T. (1998). 2-Aminopurine fluorescence studies of base stacking interactions at abasic sites in DNA: Metal-ion and base sequence effects. *Nucleic Acids Res.* **26**, 3837–3844.
38. McCullough, A. K., Dodson, M. L., Scharer, O. D., and Lloyd, R. S. (1997). The role of base flipping in damage recognition and catalysis by T4 endonuclease V. *J. Biol. Chem.* **272**, 27210–27217.
39. Lycksell, P. O., Graslund, A., Claesens, F., McLaughlin, L. W., Larsson, U., and Rigler, R. (1987). Base pair opening dynamics of a 2-aminopurine substituted Eco RI restriction sequence and its unsubstituted counterpart in oligonucleotides. *Nucleic Acids Res.* **15**, 9011–9025.
40. Bloom, L. B., Otto, M. R., Eritja, R., Reha-Krantz, L. J., Goodman, M. F., and Beechem, J. M. (1994). Pre-steady-state kinetic analysis of sequence-dependent nucleotide excision by

the 3′-exonuclease activity of bacteriophage T4 DNA polymerase. *Biochemistry* **33**, 7576–7586.
41. Xu, D., Evans, K. O., and Nordlund, T. M. (1994). Melting and premelting transitions of an oligomer measured by DNA base fluorescence and absorption. *Biochemistry* **33**, 9592–9599.
42. Rachofsky, E. L., Osman, R., and Ross, J. B. (2001). Probing structure and dynamics of DNA with 2-aminopurine: Effects of local environment on fluorescence. *Biochemistry* **40**, 946–956.
43. Rachofsky, E. L., Seibert, E., Stivers, J. T., Osman, R., and Ross, J. B. (2001). Conformation and dynamics of abasic sites in DNA investigated by time-resolved fluorescence of 2-aminopurine. *Biochemistry* **40**, 957–967.
44. Chen, L., Haushalter, K. A., Lieber, C. M., and Verdine, G. L. (2002). Direct visualization of a DNA glycosylase searching for damage. *Chem. Biol.* **9**, 345–350.
45. Parikh, S. S., Mol, C. D., Slupphaug, G., Bharati, S., Krokan, H. E., and Tainer, J. A. (1998). Base excision repair initiation revealed by crystal structures and binding kinetics of human uracil-DNA glycosylase with DNA. *EMBO J.* **17**, 5214–5226.
46. Ramstein, J., and Lavery, R. (1988). Energetic coupling between DNA bending and base pair opening. *Proc. Natl. Acad. Sci. USA* **85**, 7231–7235.
47. Fuxreiter, M., Luo, N., Jedlovszky, P., Simon, I., and Osman, R. (2002). Role of base flipping in specific recognition of damaged DNA by repair enzymes. *J. Mol. Biol.* **323**, 823–834.
48. Seibert, E., Ross, J. B., and Osman, R. (2002). Role of DNA flexibility in sequence-dependent activity of uracil DNA glycosylase. *Biochemistry* **41**, 10976–10984.
49. Verdine, G. L., and Bruner, S. D. (1997). How do DNA repair proteins locate damaged bases in the genome? *Chem. Biol.* **4**, 329–334.
50. Berdal, K. G., Johansen, R. F., and Seeberg, E. (1998). Release of normal bases from intact DNA by a native DNA repair enzyme. *EMBO J.* **17**, 363–367.
51. Kavli, B., Slupphaug, G., Mol, C. D., Arvai, A. S., Peterson, S. B., Tainer, J. A., and Krokan, H. E. (1996). Excision of cytosine and thymine from DNA by mutants of human uracil-DNA glycosylase. *EMBO J.* **15**, 3442–3447.
52. Handa, P., Acharya, N., and Varshney, U. (2002). Effects of mutations at tyrosine 66 and asparagine 123 in the active site pocket of *Escherichia* coli uracil DNA glycosylase on uracil excision from synthetic DNA oligomers: Evidence for the occurrence of long range interactions between the enzyme and substrate. *Nucleic Acids Res.* **30**, 3086–3095.
53. Xiao, G., Tordova, M., Jagadeesh, J., Drohat, A. C., Stivers, J. T., and Gilliland, G. L. (1999). Crystal structure of *Escherichia coli* uracil DNA glycosylase and its complexes with uracil and glycerol: Structure and glycosylase mechanism revisited. *Proteins* **35**, 13–24.
54. Drohat, A. C., and Stivers, J. T. (2000). *Escherichia coli* uracil DNA glycosylase: NMR characterization of the short hydrogen bond from his187 to uracil O2. *Biochemistry* **39**, 11865–11875.
55. Werner, R. M., Jiang, Y. L., Gordley, R. G., Jagadeesh, G. J., Ladner, J. E., Xiao, G., Tordova, M., Gilliland, G. L., and Stivers, J. T. (2000). Stressing-out DNA? The contribution of serine–phosphodiester interactions in catalysis by uracil DNA glycosylase. *Biochemistry* **39**, 12585–12594.
56. Stivers, J. T., and Jiang, Y. L. (2003). A mechanistic perspective on the chemistry of DNA repair glycosylases. *Chem. Rev.* **103**, 2729–2759.
57. Lloyd, R. S., Hanawalt, P. C., and Dodson, M. L. (1980). Processive action of T4 endonuclease V on ultraviolet-irradiated DNA. *Nucleic Acids Res.* **8**, 5113–5127.
58. Gruskin, E. A., and Lloyd, R. S. (1986). The DNA scanning mechanism of T4 endonuclease V. Effect of NaCl concentration on processive nicking activity. *J. Biol. Chem.* **261**, 9607–9613.
59. Lloyd, R. S. (2001). Processivity of DNA repair enzymes. *Methods Mol. Biol.* **160**, 3–14.

60. Francis, A. W., and David, S. S. (2003). *Escherichia coli* MutY and Fpg utilize a processive mechanism for target location. *Biochemistry* **42**, 801–810.
61. Bennett, S. E., Sanderson, R. J., and Mosbaugh, D. W. (1995). Processivity of *Escherichia coli* and rat liver mitochondrial uracil-DNA glycosylase is affected by NaCl concentration. *Biochemistry* **34**, 6109–6119.
62. Stanford, N. P., Szczelkun, M. D., Marko, J. F., and Halford, S. E. (2000). One- and three-dimensional pathways for proteins to reach specific DNA sites. *EMBO J.* **19**, 6546–6557.
63. Gowers, D. M., and Halford, S. E. (2003). Protein motion from nonspecific to specific DNA by three-dimensional routes aided by supercoiling. *EMBO J.* **22**, 1410–1418.
64. Jiang, Y. L., Kwon, K., and Stivers, J. T. (2001). Turning on uracil-DNA glycosylase using a pyrene nucleotide switch. *J. Biol. Chem.* **276**, 42347–42354.
65. Handa, P., Roy, S., and Varshney, U. (2001). The role of leucine 191 of *Escherichia coli* uracil DNA glycosylase in the formation of a highly stable complex with the substrate mimic, Ugi, and in uracil excision from the synthetic substrates. *J. Biol. Chem.* **276**, 17324–17331.
66. Drohat, A. C., Jagadeesh, J., Ferguson, E., and Stivers, J. T. (1999). The role of electrophilic and base catalysis in the mechanism of *Escherichia coli* uracil DNA glycosylase. *Biochemistry* **38**, 11866–11875.
67. Drohat, A. C., Xiao, G., Tordova, M., Jagadeesh, J., Pankiewicz, K. W., Watanabe, K. A., Gilliland, G. L., and Stivers, J. T. (1999). Hetronuclear NMR and crystallographic studies of wild-type and H187Q *Escherichia coli* uracil DNA glycosylase: Electrophilic catalysis of uracil expulsion by a neutral histidine 187. *Biochemistry* **38**, 11876–11886.
68. Lloyd, R. S. (1999). The initiation of DNA base excision repair of dipyrimidine photoproducts. *Prog. Nucleic Acid Res. Mol. Biol.* **62**, 155–175.
69. Bernards, A. S., Miller, J. K., Bao, K. K., and Wong, I. (2002). Flipping duplex DNA inside-out: A double base-flipping reaction mechanism by *Escherichia coli* MutY adenine glycosylase. *J. Biol. Chem.* **8**, 6.
70. House, P. G., Volk, D. E., Thiviyanathan, V., Manuel, R. C., Luxon, B. A., Gorenstein, D. G., and Lloyd, R. S. (2001). Potential double-flipping mechanism by *E. coli* MutY. *Prog. Nucleic Acid Res. Mol. Biol.* **68**, 349–364.
71. David, S. S., and Williams, S. D. (1998). Chemistry of glycosylases and endonucleases involved in base-excision repair. *Chem. Rev.* **98**, 1221–1261.
72. Guan, Y., Manuel, R. C., Arvai, A. S., Parikh, S. S., Mol, C. D., Miller, J. H., Lloyd, S., and Tainer, J. A. (1998). MutY catalytic core, mutant and bound adenine structures define specificity for DNA repair enzyme superfamily. *Nat. Struct. Biol.* **5**, 1058–1064.
73. Volk, D. E., House, P. G., Thiviyanathan, V., Luxon, B. A., Zhang, S., Lloyd, R. S., and Gorenstein, D. G. (2000). Structural similarities between MutT and the C-terminal domain of MutY. *Biochemistry* **39**, 7331–7336.
74. Noll, D. M., Gogos, A., Granek, J. A., and Clarke, N. D. (1999). The C-terminal domain of the adenine-DNA glycosylase MutY confers specificity for 8-oxoguanine.adenine mispairs and may have evolved from MutT, an 8-oxo-dGTPase. *Biochemistry* **38**, 6374–6379.
75. Kwon, K., Jiang, Y., and Stivers, J. T. (2003). Rational engineering of a DNA glycosylase specific for unnatural cytosine: pyrene base pairs. *Chem. Biol.* **10**, 351–359.
76. Jiang, Y. L., and Stivers, J. T. (2003). Novel epimerization of aromatic C-nucleosides with electron-withdrawing substituents with trifluoroacetic acid-benzenesulfonic acid using mild conditions. *Tetrahedron Lett.* **44**, 4051–4055.
77. Jiang, Y. L., and Stivers, J. T. (2003). Efficient epimerization of pyrene and other aromatic C-nucleosides with trifluoroacetic acid in dichloromethane. *Tetrahedron Letts.* **44**, 85–88.
78. Moran, S., Ren, R. X. F., Rumney, S., and Kool, E. T. (1997). Difluorotoluene, a nonpolar isostere for thymine, codes specifically and efficiently for adenine in DNA replication. *J. Amer. Chem. Soc.* **119**, 2056–2057.

79. Guckian, K. M., Morales, J. C., and Kool, E. T. (1998). Structure and base pairing properties of a replicable nonpolar isostere for deoxyadenosine. *J. Org. Chem.* **63**, 9652–9656.
80. Guckian, K. M., Krugh, T. R., and Kool, E. T. (1999). Solution structure of a replicable nonpolar base pair in DNA. *Abs. Papers Amer. Chem. Soc.* **217**, 184–ORGN.
81. Kool, E. T. (2000). Synthetically modified DNAs as substrates for polymerases. *Curr. Op. Chem. Biol.* **4**, 602–608.
82. Osman, R., Fuxreiter, M., and Luo, N. (2000). Specificity of damage recognition and catalysis of DNA repair. *Comput. Chem.* **24**, 331–339.
83. Verdine, G. L., and Norman, D. P. G. (2003). Covalent trapping of protein–DNA complexes. *Annu. Rev. Biochem.* **72**, 337–366.
84. Dhavan, G. M., Lapham, J., Yang, S., and Crothers, D. M. (1999). Decreased imino proton exchange and base-pair opening in the IHF–DNA complex measured by NMR. *J. Mol. Biol.* **288**, 659–671.
85. Palmer, A. G. (1997). Probing molecular motion by NMR. *Curr. Op. Struct. Biol.* **7**, 732–737.
86. Wyatt, M. D., Allan, J. M., Lau, A. Y., Ellenberger, T. E., and Samson, L. D. (1999). 3-methyladenine DNA glycosylases: Structure, function, and biological importance. *Bioessays* **21**, 668–676.
87. Abner, C. W., Lau, A. Y., Ellenberger, T., and Bloom, L. B. (2001). Base excision and DNA binding activities of human alkyladenine DNA glycosylase are sensitive to the base paired with a lesion. *J. Biol. Chem.* **276**, 13379–13387.
88. Allan, B. W., and Reich, N. O. (1996). Targeted base stacking disruption by the EcoRI DNA methyltransferase. *Biochemistry* **35**, 14757–14762.
89. Krosky, D. J., Schwarz, F. P., and Stivers, J. T. (2004). Linear-free energy correlations for enzymatic base flipping: How do damaged base pairs facilitate specific recognition? *Biochemistry*, in press.

Bacteriophage T2Dam and T4Dam DNA-[N6-adenine]-methyltransferases[1]

STANLEY HATTMAN* AND
ERNST G. MALYGIN[†]

*Department of Biology, University of
Rochester, Rochester, NY 14627–0211
[†]Institute of Molecular Biology, State
Research Center of Virology and
Biotechnology, Novosibirsk 630559, Russia

I. Historical Background	68
A. Host-Induced Modification of Phage T2: Loss of Phage DNA Glucosylation	68
B. Phage DNA Methylation In Vivo	70
C. Altered Phage DNA MTases	71
D. Comparison of T4Dam Orthologs	72
II. Binding Properties of T4Dam	73
A. Subunit Structure: Dependence on Interaction with Substrate Oligodeoxynucleotide Duplex	73
B. Interaction with ODN Duplexes Containing Native or Altered Recognition Sites	76
III. Kinetic Properties of T4Dam Methylation of Substrate Duplexes Containing Native or Altered Recognition Sites	84
A. Steady State Analysis	84
B. Pre-Steady State "Burst" Analysis	86
C. Steady State Mechanism: Kinetic Evidence for a Catalytically Essential Conformational Change in the Ternary Complex	92
D. Processivity and Orientation to the Methylation Target	99
E. Single Turnover Analysis	105
F. Comparison to a DNA-[N4-cytosine]MTase	109
IV. Structure of T4Dam	112
A. Alignment of Motifs and Secondary Structures	112
B. Three-Dimensional Structure from X-ray Crystallography	113
V. Concluding Comments	118
References	120

DNA methyltransferases (MTases)[1] are important enzymes that methylate DNA as a post replicative event. DNA MTases, encoded by both cellular and viral genes, catalyze methyl group transfer from S-adenosyl-L-methionine

---

[1] Abbreviations: AdoHcy, S-adenosyl-L-homocysteine; AdoMet, S-adenosyl-L-methionine; 2-AP, 2-aminopurine; hmC, 5-hydroxymethylcytosine; m6A, N6-methyladenine; MTase, methyltransferase; ODN, oligodeoxynucleotide; TRD, target-recognition-domain.

(AdoMet), producing S-adenosyl-L-homocysteine (AdoHcy) and methylated DNA. The methyl group acceptor atom is either an exocyclic amino nitrogen (N6-Ade or N4-Cyt) or a ring carbon (C5-Cyt). The DNA-[amino]-MTases [EC 2.1.1.72 and 113] transfer methyl groups directly to the exocyclic nitrogen without the formation of a covalent enzyme-DNA intermediate, which occurs with the C5-Cyt MTases [EC 2.1.1.73]. While most prokaryote DNA MTases are components of restriction-modification systems important in protecting cells from foreign DNAs, certain MTases do not have cognate restriction enzymes associated with them. These include a family (Dam) of prokaryotic DNA-adenine MTases that methylate Ade in GATC sequences. Several bacteriophages, such as T2 and T4, also encode Dam MTases. Generally speaking, the Dam MTases are not essential for viability of bacteria or phage; however, they do have a variety of functions including regulation of transcription of certain genes, timing of DNA replication initiation, and protection against restriction endonucleases, and they play a crucial role in pathogenicity of intestinal bacteria. Because they methylate specific nucleotide sequences, they provide excellent objects for studies on protein–DNA interactions. Valuable insights into the organization/function of MTases have come from the identification of common motifs discovered by amino acid-sequence alignments. The solution of a number of MTase crystal structures has added key details on the specific protein–DNA and protein–cofactor interactions. However, genetic and biochemical analyses are essential for a more complete understanding of the functioning of these enzymes. Here, we present results from all three approaches directed at characterizing the bacteriophage T2/T4Dam DNA-[N6-adenine]-MTase.

I. Historical Background

A. Host-Induced Modification of Phage T2: Loss of Phage DNA Glucosylation

The second half of the twentieth century witnessed a revolution in genetics that grew out of pioneering research with prokaryotes and lower eukaryotes. Among the earliest players in this saga were *Escherichia coli*, *Salmonella typhimurium*, and their bacteriophages. Although phage work as we knew it has virtually disappeared, for several decades it was at the forefront of molecular biological research. The beginning of our story goes back more than 50 years to when two reports (1, 2) appeared that described the phenomenon of host-induced modification or host-controlled variation. Since our chapter focuses on the Dam DNA-[N6-adenine] MTases of phages T2 and T4, we describe results related only to this system (1). *E. coli* B variants that had gained (adsorption) resistance to phage T4 were isolated and these variants fell into two phenotypic classes: one was resistant to T4 only [=B/4] while the second was resistant to phages T3, T4, and T7 [=B/3,4,7]. When the latter class was used as a host for infection with phage T2, the story took an unexpected twist. First, T2 could not make plaques on B/3,4,7. However, in a classical one-step growth experiment, T2 adsorbed to these cells and lysed them at the end of a normal latent period. But, there did not appear to be any significant amount of new phage production, as shown by plaque-forming unit assays using *E. coli* B as the indicator host. The related enteric strain, *Shigella dysenteriae* strain Sh, was known to be a good host for T phages; i.e., T2 propagated on either B or Sh for one or multiple growth cycles had a similar efficiency of plating on both hosts. When Sh was used as the indicator host, an almost normal burst of T2 progeny phage was found after a single growth cycle in B/3,4,7, but they had an efficiency of plating of only 10^{-3}–10^{-4} on B relative to Sh. This indicated that an alteration in viral host range had resulted during a single growth cycle in B/3,4,7 (this did not occur in the B/4 strain). These results illustrate the new phenomenon of host-induced modification. Thus, *S. dysenteriae* Sh is a permissive host for the modified progeny phage, designated T°2, while *E. coli* B is a nonpermissive (or restricting) host and B/3,4,7 (renamed B/4$_o$) is a modifying host. The same modification was observed to occur with phage T6 infection of B/4$_o$.

The molecular basis for these results remained a mystery for more than a decade. During that period, new and surprising information was obtained about the chemistry of DNA in general and that of phages T2, T4, and T6 (the so-called "T-even" phages) in particular. First and foremost, the Watson-Crick structure of DNA was published in 1954. Then, it was discovered that T-even phage DNA lacks cytosine, which is completely replaced by 5-hydroxymethyl-Cyt (5-hmC) (3); moreover, these residues are glucosylated and the nature and extents of glucosylation were phage-specific (4). The replacement of C by hmC occurs as the result of the action of three phage-encoded enzymes, which dephosphorylate dCTP to dCMP, hydroxymethylate it to hmdCMP, and then phosphorylate it to hmdCTP, respectively (reviewed in (5)).

Thus, replacement of C by hmC occurs at the level of nucleotide pool metabolism, prior to DNA synthesis. In contrast, hmC-glucosylation is a postreplicative event mediated by phage-specific enzymes, glucosyl transferases, which utilize a host metabolite, uridine diphosphoglucose (UDPG), as the glucosyl donor. The connection between all these observations and the T°2 story was solved in 1963 (6). First, the *E. coli* B/4$_o$ strain was shown to be a *gal*$^-$ mutant lacking UDPG-pyrophosphorylase activity, an enzyme required for the synthesis of UDPG. Consequently, production of phage in this host results in progeny T°2 DNA lacking glucose (6). Finally, when a spontaneous *gal*$^+$ revertant from B/4$_o$ *gal*$^-$ was isolated, it produced normal T2 instead of modified T°2 phage. The alteration in T°2 host range was then accounted for by the observation that unglucosylated T*2 DNA was rapidly degraded to acid soluble material following infection of restricting, but not permissive, hosts (7).

The T*2 story represented the first case where the chemical basis of a host-induced modification was elucidated; it would take another few years before it was shown that DNA methylation is involved in the more general host-specificity determination (8–10).

B. Phage DNA Methylation *In Vivo*

Understanding T2 ("sweet") vs T*2 ("sour") phages enabled us to devise strategies to screen for mutants (gt^-) defective in glucosyl *t*ransferase activity (reviewed in (*11*)) as well as for (rgl^-) host cell mutants defective in *r*estriction of *gl*ucoseless T-even phage DNA (reviewed in (*12*)). Unlike wild type gt^+ T-even phage, nonglucosylated gt^- mutants are strongly restricted on hosts lysogenic for P1 prophage (*13*). However, it was possible to obtain from phages T2 gt^- and T4 gt^- (but not T6 gt^-) derivatives capable of growing on P1 lysogens (*13*). Since these were still gt^- (unable to grow on *E. coli* B), they were designated *uP1* (*u*nrestricted by *P1*), in contrast to their *rP1* parents (*r*estricted by *P1*).

A key portion of the solution to the puzzle came from the knowledge that phage T2 and T4 (but not T6) virion DNAs contain 6-methylaminopurine (= N6-methylA = m6A) (*14, 15*). Moreover, infection with T2 or T4 (but not T6) induced an increase in DNA-[N6-adenine]-methyltransferase (MTase) activity (*16–18*). In these experiments, cell-free extracts were prepared from cultures at different times postinfection, and then assayed for DNA MTase activity *in vitro* by monitoring the transfer of radioactively labeled methyl groups from donor S-adenosyl-L-methionine (AdoMet) to an acceptor DNA. This was determined by monitoring the conversion of the AdoMet-radioactive label to an acid insoluble form, since the reaction products were methylated DNA and unlabeled S-adenosyl-L-homocysteine (AdoHcy).

The final connection between these observations and the *rP1/uP1* story became evident through a systematic comparison of the m6A contents of virion DNAs (*19*). In these studies, phage DNA was labeled during infection of permissive cells in medium containing [^{3}H]-2-adenine. After purifying progeny phage, the virion DNA was extracted, acid hydrolyzed, and the resulting purine bases separated by paper chromatography. It was immediately clear that both the T2 gt^- *uP1* and T4 gt^- *uP1* virion DNAs were hypermethylated compared to their respective parental gt^- *rP1* forms. This suggested that resistance to P1 restriction might be afforded by the additional methylation; this is, in fact, the case and will be discussed in more detail in the following text. It was also interesting that T2 gt^- and T4 gt^- phage each had higher m6A contents than their respective gt^+ parents, suggesting that DNA glucosylation interfered with *in vivo* DNA methylation. Lastly, the various T2 forms always exhibited higher (~50%) m6A contents than the corresponding T4 forms, indicating that the T2 MTase methylated virion DNAs to higher extents than the T4 MTase (although other explanations were also tenable); this issue will be revisited in Section C.

Through laborious screening of many individual gt^- $uP1$ plaques, variants were isolated that had regained sensitivity to P1 restriction; these were designated gt^- u^RP1 (20). Surprisingly, virion DNA from gt^- u^RP1 lacked m6A. Genetic studies then showed that independently isolated u^RP1 mutations mapped close to and on both sides of the original $uP1$ mutation, suggesting that the different u^RP1 mutations were alleles of the same gene, presumably encoding the DNA MTase. Therefore, a new nomenclature was adopted. The $rP1$ genotype was replaced by dam^+ to indicate the wild-type Dam MTase gene, $uP1$ was replaced by dam^h (encoding the Dam^h MTase), and u^RP1 was replaced by dam^h dam-x (encoding the Dam^h Dam-x MTase). Finally, the fact that T6 gt^- never yielded a $uP1$ (dam^h) derivative could be attributed to its lack of a functional Dam MTase. This was supported later by Southern hybridization studies, which indicated that T6 does not have a dam gene homolog (21).

C. Altered Phage DNA MTases

Bacterial strain-specific DNA MTases were discovered in the early 1960s (reviewed in (22)). Thus, with the isolation of gt^- $uP1$ mutants resistant to P1 restriction, nature had provided dam^h mutations in the phage DNA MTase, allowing them to hypermethylate virion DNA. [*Note:* The proper designation of the phage MTase is M.EcoT2Dam; however, it will be referred to hereafter with the common name T2Dam.] The wild type T2Dam and mutant T2Dam^h enzymic forms were partially purified from infected *E. coli*, and *in vitro* methylation analyses were carried out using various DNA substrates (23). The T2Dam^h mutant form exhibited a relatively higher thermal lability than the T2Dam$^+$ enzyme, in both the presence and absence of substrate methyl donor AdoMet. Using various polymeric hmCyt-containing DNA substrates, the T2Dam^h MTase showed a reduced K_m (for DNA) and reached higher extents of methylation relative to the T2Dam$^+$ enzyme; in contrast, the two enzymes showed similar methylation of nonviral, cytosine-containing DNAs (23). The *in vitro* specificity of the two enzymes was derived from determining the sequences of [^{14}C-methyl]-labeled oligodeoxynucleotides (ODNs) generated following methylation, enzymatic digestion, two-dimensional paper electrophoresis and autoradiography (24). Despite the two- to three-fold higher extents of (unglucosylated, unmethylated hmCyt-phage) DNA methylation attained with the Dam^h MTase, it was not possible to discern any qualitative differences in the labeled ODN species; both enzymes methylated sites consistent with a recognition sequence of G-A-Y [where Y is C or T]. This study was extended to the analysis of *in vitro* methylation phage λ DNA (25).

Again, the T2Dam^h enzyme gave a higher extent of DNA methylation, and only T2Dam^h methylation protected λ against P1-restriction (assayed biologically by spheroplast transfections). In the same study, dinucleotide analysis of the nearest neighbors to *in vitro* [^{14}C-methyl] labeled m6A showed that the Dam^h MTase methylated GAC sequences much more readily than did the Dam MTase. These results were consistent with the notion that the Dam^h MTase protected λ DNA by methylating the P1 recognition sequence to A-G-m6A-C-C (26).

The *in vitro* experiments previously described were all done at high enzyme-to-DNA ratios, a condition that is not likely to apply *in vivo*, where the *dam*$^+$ virion DNA is methylated primarily in the palindromic tetranucleotide sequence, GATC (a subset of GAY). GATC is referred to as the canonical recognition (or target) sequence (27). In fact, complete methylation of the GATC sites accounts for all of the m6A in T4 *dam*$^+$ virion DNA. Thus, it follows that the higher methylation levels present in virion DNA from T2 *dam*$^+$, as well as T2 (and T4) *dam*^{h}, must be due to noncanonical GAY site methylation. In retrospect, we would have observed qualitative differences in the *in vitro* methylation patterns had we chosen to compare the T4Dam and Dam^h MTases rather than those of T2.

The nature of the *dam*^{h} mutation (in both T2 and T4) was shown later to be a single base substitution that alters the Pro126 residue to Ser (28). Under steady state conditions, the T2Dam^h and Dam MTase forms exhibited similar k_{cat} values with synthetic 24mer ODN duplexes containing the canonical GATC/GATC site; and these values were ~1.5-fold higher than for the T4Dam MTase (29). The largest difference in methylation capability among the various Dam MTase forms was observed with a duplex containing a noncanonical asymmetrical GACT/AGTC site; the k_{cat} for T2Dam^h was 4-fold higher than T2Dam and 7-fold higher than T4Dam. The differences between the T2 and T4 enzymes at the level of amino acid sequence will be discussed in Section D.

D. Comparison of T4Dam Orthologs

DNA-[N6-adenine]-MTases are distributed among more than 100 enteric bacteria and frequently encoded by their phages, including T1 (30), P1 (31), and T2 and T4. In addition, a gram-positive organism, *Streptococcus* (*Diplococcus*) *pneumoniae*, encodes such DNA MTases, including M.DpnA and M.DpnM in the DpnII restriction system (32), as do certain viruses infecting a eukaryotic *Chlorella*-like green alga (33). The single most important advance in the study of the phage Dam MTases was the cloning and sequencing of the T4 *dam* gene (34, 35). The *E. coli* DNA MTase, EcoDam, was one of the first to have been studied in detail; it symmetrically methylates the A in the palindromic sequence, GATC (26). In contrast to the phage enzymes, EcoDam has a much higher degree of selectivity, being unable to methylate noncanonical sites. Moreover, it is unable to methylate hmCyt-containing phage DNA; this was a fortunate circumstance since it allowed us to study *in vivo* DNA methylation by the phage-encoded MTases without having to be concerned with the action of the host enzyme. One of the earliest comparative analyses of MTases was made for T4 viral and three bacterial *dam* genes (36). Despite the lack of DNA sequence similarity, a computer alignment identified "patch homology" at the amino acid sequence level between T4Dam, EcoDam, an M.DpnM. T4 and EcoDam share four regions of similarity, designated I to IV, ranging from 11 to 33 residues and with 45 to 64% identity. Regions I, III, and IV are also found in M.DpnM (36a). Later analysis showed that a (D/N/S)-P-P-(F/Y) motif in homology region IV is highly conserved among all DNA-[amino] MTases (37, 38).

Using a labeled T4 *dam* gene fragment, genomic digests of phage T2 and T6 virion DNAs were probed by Southern hybridization (21). While a T2 *dam* gene homolog was readily observed, no hybridization with T6 could be detected. This was consistent with the absence of T6 virion DNA methylation, suggesting that T6 lacked a homologous *dam* gene. The T2 *dam* gene was cloned, sequenced, and compared to T4 *dam* (21). While there were 22 nucleotide differences between T2 and T4, there were only three amino acid differences, viz., (T4 Ser20, T2 Pro), (T4 Asn26, T2 Asp), and (T4 Asp188, T2 Glu). A cloned T4 *dam* gene was altered by site–directed mutagenesis at these three sites (39), and the mutant MTases they encoded were purified to homogeneity, as for the wild-type enzyme (40). *In vitro* kinetic analyses showed that while they exhibited similar K_m values for both donor AdoMet and substrate DNA, T2Dam had a 2- to 3-fold higher k_{cat} than T4Dam (39). Moreover, a single change of T4Dam residue 20 or 26 to the one found in T2Dam was sufficient to increase the k_{cat} to that characteristic of T2Dam (39). This could account for the higher *in vivo* methylation level of T2 gt^- virion DNA compared to T4 gt^- (19), presumably due to noncanonical site methylation. To investigate this further, we also compared the intracellular enzyme levels following phage infection. Results from Western blots showed that the same amounts of MTase protein were produced after infection with T2 and T4. Thus, the different levels of virion DNA methylation could be attributed to the higher rate of T2Dam methylation and its ability to methylate noncanonical sites.

II. Binding Properties of T4Dam

A. Subunit Structure: Dependence on Interaction with Substrate Oligodeoxynucleotide Duplex

Based on glycerol gradient centrifugation and gel filtration analyses under native conditions, T4Dam was calculated to have a molecular weight of 30,690 (40) and an apparent molecular mass of 30.0 kDa from SDS-polyacrylamide gel electrophoresis analysis. These values were in good agreement with 30.4 kDa from the deduced amino acid sequence. From these results, one could infer that T4Dam is a monomer in solution. However, it is known that an enzyme's subunit state may depend on the protein concentration as well as on the presence or absence of substrate(s) or allosteric effector (41). Therefore, we also investigated whether the presence of substrate synthetic ODN duplexes influenced the apparent T4Dam subunit structure (42). For this purpose, we first analyzed the interaction between highly purified T4Dam with defined duplexes made by annealing various ODNs:

duplex I, from palindromic 32mer 5' CGGGTACCCTATTGGATCCAAT-AGGGTACCCG 3'

duplex II, from complementary 20mers
  5' GTGAAATGGATCCTAAACTG 3' and 3' CACTTTACCTAGGA-TTTGAC 5'

duplex III, from palindromic 12mer 5' CGCGGATCCGCG 3'.

The apparent molecular weight of the free enzyme was reexamined by gel filtration and sedimentation; the resulting calculated values of 28.4 and 28 kDa, respectively, were in fairly good agreement with the original data (40). However, in the presence of either duplex I or duplex II (the latter is termed the 20mer specific duplex), there was a significant increase in the size of the complex, corresponding to two enzyme molecules per duplex (42). In contrast, the short [12mer] duplex III bound only a single enzyme molecule. This might be due to a lower affinity for the shorter duplex or to instability of an E_2S complex. Alternatively, the longer duplexes might accommodate a side-by-side nonspecific binding of two enzyme molecules along the length of the duplex, which would be sterically impossible with the 12mer duplex. In any event, these results showed that an E_2S complex can be formed *in vitro*. Subsequent analyses revealed that the E_2S complex was catalytically functional (43). Recently, an X-ray crystal structure of a T4Dam-DNA-AdoHcy complex (44) was shown to have an enzyme/DNA (a 12mer duplex) ratio of 2:1.

To study in more detail the conditions for the apparent oligomerization of T4Dam, we took advantage of the ability of glutaraldehyde to cross-link protein subunits (through covalent bonds between the epsilon amino residues of closely spaced lysine residues (45)). Cross-linking stabilizes preexisting protein oligomers and allows their visualization by SDS-polyacrylamide gel electrophoresis. To reduce possible nonspecific cross-linking, we chose concentrations of 1 μM T4Dam and 0.0025% glutaraldehyde; these concentrations were substantially lower than those commonly used in such experiments. Sodium borohydride was added to stabilize the cross-links, and the samples were analyzed by SDS-polyacrylamide gel electrophoresis (Fig. 1).

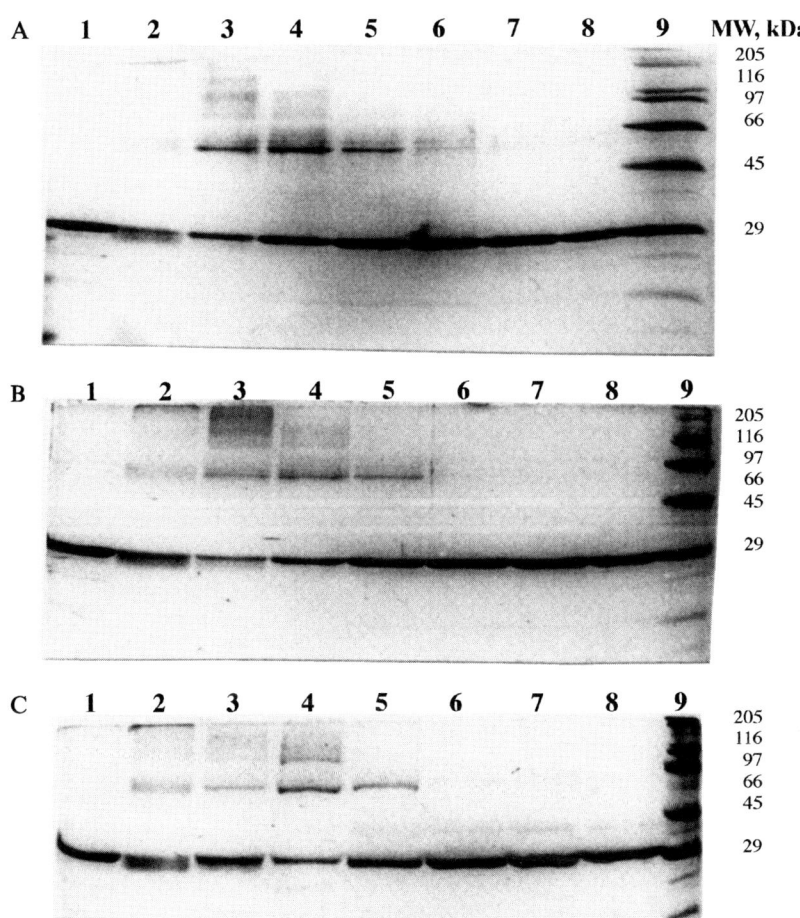

FIG. 1. Glutaraldehyde cross-linking of T4Dam (1 μM) in the presence of varying concentrations of duplex: (A) duplex N/A; (B) duplex N/M (structures of these duplexes are given in Section II.B.2). Lane 1, T4Dam, untreated; lane 2, T4Dam treated with 0.0025% glutaraldehyde; lanes 3–8, T4Dam treated in the presence of duplex at concentrations of 0.25, 0.5, 1.0, 2.0, 4.0, and 8.0 μM, respectively; lane 9, molecular weight standards. (C) Cross-linking in the presence of 5 μM AdoMet. Lane 1, T4Dam, untreated; lane 2, T4Dam + AdoMet, treated with 0.0025% glutaraldehyde; lanes 3–8, T4Dam + AdoMet, treated in the presence of the N/M duplex at concentrations of 0.1, 0.25, 0.5, 1.0, 2.0, and 4.0 μM, respectively; lane 9, molecular weight standards. Taken from (45) by permission of Oxford University Press.

Glutaraldehyde treatment of free T4Dam led to the formation of some SDS-resistant oligomers of variable size (45). However, in the presence of a specific 20mer duplex and at a molar ratio of [enzyme]/[duplex] = 4, the predominant oligomeric form was that of an E_2S complex, consistent with our previous results (42). As the [enzyme]/[duplex] ratio was decreased, the formation of cross-linked oligomeric decreased. Similar results were observed if the enzyme had been preincubated with AdoMet prior to the addition of the duplex (Fig. 1B). These experiments allowed us to adjust reaction conditions where T4Dam exists primarily as a monomer ([enzyme] ≪ [duplex]) or as a dimer ([enzyme] ≫ [duplex]), so that we could compare the catalytic activities of these forms (see Section III.E). It seems that chemical cross-linking at different [enzyme]/[duplex] ratios could be useful for other DNA MTases to determine possible changes in their subunit states.

B. Interaction with ODN Duplexes Containing Native or Altered Recognition Sites

1. Gel Shift Analysis

The catalytic mechanism for methyl transfer for the DNA-[C5-cytosine] MTases involves covalent bonding of an MTase-cysteine sulfur to the C6 ring carbon of the target cytosine residue (46). This is followed by C5-activation to react with an electrophile, the methyl group of AdoMet. In contrast, because their target is an exocyclic NH_2 group, the [N6-adenine] and [N4-cytosine] MTases must utilize a different mechanism for methyl transfer. Not unexpectedly, the DNA-[amino]-MTases are more closely related to one another than to the DNA-[C5-cytosine]-MTases (37). Therefore, elucidating the mechanism of action of DNA-[amino]-MTases was of great interest. In this regard, we examined the effect of base modifications in the GATC target site on T4Dam complex formation. For this purpose, we employed defined synthetic 24mer ODNs containing a base analog that potentially alters (or removes) a single contact. All ODN duplexes studied had the same general sequence (shown in the following text), where derivatives contained individual A or G substitutions in the specific GATC target site (underlined). The substitutions were introduced in either the upper or lower strand: G → 2-aminopurine (2-AP; G → N), G → 7-deazaguanine (G → Z), A → 2-AP (A → N), or A → purine (A → P). In order to be able to monitor the binding, the ODNs were radioactively end-labeled using [^{32}P-γ-ATP] and T4 polynucleotide kinase, and the complexes were separated by nondenaturing polyacrylamide gel electrophoresis.

5'- CGCGGGCGGCG<u>GATC</u>CGGGCGGGC -3'   [upper strand]
3'- GCGCCCGCCGC<u>CTAG</u>GCCCGCCCG -5'   [lower strand]

While the common flanking sequence for duplexes 1–5 differed from that for duplexes 6–15 (Table I), the difference in flanking sequences did not strongly influence complex formation. Also, substitutions introduced in either the upper or lower strand produced the same effect, so only the modifications in the upper strand are presented in Table I.

Both A → P and A → N substitutions remove a potential contact point (adenine-N6) with T4Dam and one of the two major groove Watson-Crick H-bonds; the A → N substitution, however, introduces an H-bond in the minor groove (47). These substitutions did not greatly reduce binding capability of the modified substrates; i.e., the K_d increased from 2- to 6-fold, with the A → N substitution binding poorer relative to the A → P substituted duplex. As for the canonical duplex I, presence of AdoMet enhanced binding 2- to 3-fold for both substitutions. The G → N substitution deletes an O6 ketogroup, which normally participates in H-bonding, while the G → Z substitution replaces the imidazole ring N7 with a carbon, but that does not perturb H-bonding. Both the G → N and G → Z substitutions decreased complex formation circa 40- to 50-fold compared with native duplex I. These results were consistent with the G residue playing a critical role in T4Dam binding.

Since a natural DNA may contain some structural defects, such as nicks, abasic gaps, deletions, or chemical modification of nucleotide residues, it was important to know how these alterations might influence the methylation process. Thus, we used self-complementary 12mer and 20mer synthetic ODN duplexes that had either a normal, or defective or methylated target site (some examples are presented in Table I); in addition, the effect of adding/omitting substrate AdoMet was examined (48, 49). The results of these studies are summarized as follows: (i) T4Dam bound with approximately 100-fold higher affinity to the 20mer specific (GATC-containing) duplex containing the canonical palindromic methylation seqence GATC, than to a duplex containing another palindrome, GTAC. (ii) No stable complex was formed with a synthetic 12mer specific (GATC-containing) duplex, although T4Dam could methylate it. This indicates that there was no correlation between formation of a catalytically competent 12mer-T4Dam complex and one stable to gel electrophoresis. (iii) Formation of a stable complex did not require that both strands be contiguous or completely complementary. Some target defects led to sharp decreases in stability of the complexes, while others were neutral or even increased affinity for the enzyme. A duplex lacking either T or AT formed a strong complex, whereas a weak complex was produced when there was a nucleotide gap of C or G within GATC. Thus, having only one-half of the recognition site intact on one strand was sufficient for stable complex formation provided that the 5′G · C base pairs were present at both ends of the palindrome. Since absence of either a G or C abolished T4Dam binding, we concluded that T4Dam recognizes both strands. (iv) Absence of a single internucleotide

TABLE I
Characteristics of Binding and Methylation of Duplexes with a Native or Modified Recognition Site by T4Dam

Duplex no[b]	Recognition site		In absence of AdoMet	In presence of AdoMet					
	Structure[a]	Comment	K_d, nM[c]	K_d, nM	$\Delta\Delta G^\circ_{bind}$,[d] kcal/mol	K_m,[c] nM	k_{cat}, $s^{-1} \times 10^{-3}$	k_{cat}/K_m, $M^{-1} s^{-1} \times 10^{-6}$	k_{cat}/K_m,[d] relative
0	-C-T-A-G- / -G-A-T-C-	inverted sequence	>1500	>2500	>2.9		no methylation		
1	**-G-A-T-C-** / **-C-T-A-G-**	**canonical sequence**	**54 (7)**	**18 (3)**	**0.0**	**7.7 (2.1)**	**14 (3.0)**	**1.80**	**1.0**
2	-G-P-T-C- / -C-T-A-G-	N6 amino group deleted	105 (4)	47 (2)	0.5	2.5 (1.2)	0.99 (0.15)	0.40	0.22
3	-G-N-T-C- / -C-T-A-G-	N6 amino group deleted	256 (8)	95 (3)	0.9	8.7 (3.0)	2.6 (0.5)	0.30	0.16
4	-N-A-T-C- / -C-T-A-G-	O6 keto group deleted	1700 (85)	1480 (80)	2.6	108 (50)	5.7 (3.7)	0.05	0.03
5	-Z-A-T-C- / -C-T-A-G-	N7 imidazole atom deleted	1450 (150)	935 (90)	2.3	12.3 (5.7)	13 (3.0)	1.10	0.58
6	**-G-A-T-C-** / **-C-T-A-G-**	**canonical sequence**	**43 (6)**	**17 (4)**	**0.0**	**6.3 (1.5)**	**15 (1.0)**	**2.38**	**1.0**
6m	-G-M-T-C- / -C-T-A-G-	N6 methyl group added	23 (3)	8.5 (2)	−0.40	6.0 (1.3)	10 (1.0)	1.67	0.70
7	-G-A-T-C- / -C-^T-A-G-	phosphate deleted	>1000	>1000	>2.3	10.2 (2.1)	14 (1.0)	1.27	0.53

#	Sequence 1	Sequence 2	Description	Col5	Col6	Col7	Col8	Col9	Col10	Col11
8	-G-A-T-C-	-C-T^A-G-	phosphate deleted	28 (6)	9 (2.3)	−0.37	24.6 (2.3)	5.8 (0.3)	0.24	0.10
9	-G-A-T-C-	-C-T-A^G-	phosphate deleted	18 (5)	5 (2)	−0.71	6.2 (1.2)	1.8 (0.2)	0.21	0.09
10	-G-A-T-C-	(·) T-A-G-	C residue gap	>1450	>1450	>2.5	27.7 (3.7)	24 (1.0)	0.76	0.32
11	-G-A-T-C-	(··) A-G-	TC residues gap	>1000	>1000	>2.3			very low methylation	
12	-G-A-T-C-	-C(·) A-G-	T residue gap	35 (17)	12 (10)	−0.20			very low methylation	
13	-G-A-T-C	-C(··) G-	AT residues gap	16 (5)	2.7 (0.5)	−1.1			no methylation	
14	-G-A-T-C-	-C-T-A (·)	G residue gap	≫1450	≫1450	>2.5	71 (11)	2.9 0.2	0.035	0.015
15	-G-A-T-C-	-C-A-T-G-	double mismatch	156 (45)	3.3 (1.7)	−0.95	5.3 (1.1)	0.5 (0.1)	0.06	0.025

[a]P = purine; N = 2-aminopurine; Z = 7-deazaguanine; M = N^6-methyladenine; ^ = absence of a phosphate; (·) = absence of a nucleotide.
[b]Duplexes 0–5 and 6–15, respectively, had identical sequences flanking the target site within each set but were different between sets.
[c]K_d values were obtained by the gel shift assay (48, 51). Steady state kinetic parameters were determined in (51, 64). Standard deviations are in parentheses.
[d]Values of $\Delta\Delta G^{\circ}_{bind}$, and relative k_{cat}/K_m for duplexes 1–5 were calculated relative to duplex 1, while those for duplexes 6–15 were calculated relative to duplex 6. The actual values of $\Delta\Delta G^{\circ}_{bind}$, and k_{cat}/K_m for duplexes 1 and 6 were close to one another.

phosphate strongly reduced complex formation only when missing between the T and C residues. This suggests that if T4Dam makes critical contact(s) with a backbone phosphate(s), then the one between T and C is the only likely candidate. (v) Compared to the unmethylated, specific ODN, the hemimethylated 20mer specific duplex had a slightly increased (about 2-fold) ability to form complexes with T4Dam. In contrast, hemimethylation had variable effects on defective duplexes; i.e., there was no correlation between the effect of hemimethylation with the nature/location of recognition site defect. (vi) T2Dam formed complexes with the hemimethylated 20mer specific ODN relatively poorly compared to T4Dam, suggesting that the T2 enzyme had a lower affinity for this substrate. This indicated that there is no direct correlation between MTase catalytic activity and complex formation/stability.

The cost of modified duplex recognition by the EcoRI endonuclease, for which X-ray crystallographic structures were known, was studied in detail (50). It was possible to calculate the difference in standard binding free energy between the canonical site and each modified site, $\Delta\Delta G^o_{bind} = -RT\ln(K_d/K_{d\,mod})$. These values may include a variety of changes in interactions, including protein–base ($\Delta\Delta G^o_{base}$), protein–phosphate ($\Delta\Delta^o_{phos}$), and a more general "reorganization" term ($\Delta\Delta G^o_{reorg}$), for changes in conformational contribution to the interactions. From a number of lines of evidence, and in comparison to the X-ray structure of the EcoRI–DNA complex, the energetic contribution of a single protein–base hydrogen bond to binding free energy was calculated to be about -1.4 kcal/mol.

We applied a similar analysis to evaluating the energetic consequences of different purine modifications on T4Dam interaction with defective/modified duplexes. We found only small penalties in $\Delta\Delta G^o_{bind}$ value (from circa $+0.4$ to $+1.1$ kcal/mol) for A $\to$ P or A $\to$ N, each of which removes a potential T4Dam contact point (A-N6) (51). These small penalties are likely the net of an unfavorable contribution for loss of the hydrogen bond to A-N6 (estimated at $+1.4$ kcal/mol (50)) and a compensating favorable effect on the reorganization energy ($\Delta\Delta G^o_{reorg}$). Removal of a major groove Watson-Crick constraint may facilitate achieving the precise local DNA structural features required for maximum protein–DNA complementarity at the interface. The addition of a 2-amino group in the A $\to$ N substitution introduces a Watson-Crick constraint in the minor groove; this might account for the slightly greater penalty ($\sim+1.0$ kcal/mol) observed than that for the A $\to$ P substitution ($\sim+0.5$ kcal/mol). For A substitutions, the penalties were the same in the presence or absence of AdoMet; thus, enhancement of complex formation by AdoMet is not related to any direct influence on T4Dam interaction with the A residue.

In contrast, we found significantly greater binding free energy penalties ($\Delta\Delta G^o_{bind}$ of circa $+1.9$ to $+2.6$ kcal/mol) for the G to N and Z substitutions (51). These penalties could result from changes in protein–phosphate contacts

($\Delta\Delta G^o_{phos}$) and/or conformational factors ($\Delta\Delta G^o_{reorg}$), in addition to the removal of the potential contact points O6 or N7 ($\Delta\Delta G^o_{base}$). Some independent evidence that could help assess whether, in addition to protein–base contacts, protein–phosphate contacts have also been perturbed is the value of $\Delta\Delta G^o_{bind} > 2.3$ kcal/mol for duplex 7 (Table I) with deleted internucleotide phosphate between T and C residues in the bottom chain. On the other hand, elimination of the T residue in duplex 12 followed by some binding free energy gain and elimination of both A and T residues (duplex 13) is accompanied by a more significant advantage in free energy of binding (−1.1 kcal/mol). Evidently, here is a compensating favorable effect on the reorganization energy ($\Delta\Delta G^o_{reorg}$).

2. Fluorescence Analysis: Evidence for "Base Flipping" and a Dual Role for Substrate AdoMet

Three-dimensional structures of several DNA-[Cyt-5]-MTases have been described (52–55). A most surprising and exciting finding was that the C residue to be methylated was "flipped out" of the DNA helix (54, 55). We have taken advantage of alternative methodologies (56–62) to study base flipping by T4Dam (45). One of these methods is based on the substitution of a target A by 2-AP, which serves as a fluorescent probe. It is generally accepted that N introduced in the target-base position of a specific ODN duplex mimics the behavior of the A residue (56, 57, 61, 62). The fluorescence intensity of N incorporated into double-stranded DNA is very low, but sharply increases if the base (or nucleoside) is flipped out into the cavity of the enzyme's active site. Synthetic 20mer ODN duplexes containing 2-AP (A → N substitution) and m6A (A → M substitution) in the GATC target site (underlined) were used as DNA substrates:

5'- CAGTTTAGG<u>GNTC</u>CATTTCAC - 3'   **N/A** duplex
3'- GTCAAATCC<u>CTAG</u>GTAAAGTC - 5'

5'- CAGTTTAGG<u>GNTC</u>CATTTCAC - 3'   **N/M** duplex(hemimethylated)
3'- GTCAAATCC<u>CTM</u>GGTAAAGTC - 5'

5'- CAGTTTAGG<u>GNTC</u>CATTTCAC - 3'   **N/N** duplex(double substitution)
3'- GTCAAATCC<u>CTN</u>GGTAAAGTC - 5'

As shown in Fig. 2, addition of T4Dam to duplex N/A resulted in an increase in 2-AP fluorescence. The intensity was approximately proportional to the duplex concentration, and exhibited a 50-fold signal increase at saturating enzyme concentration. These results were analogous to those observed with several other MTases (56, 57, 60–62) and consistent with flipping of the 2-AP residue out of the DNA helix. With increasing T4Dam concentration ([enzyme]/[duplex] ratios >1), we observed a decrease in the fluorescence intensity. This was attributed to the formation of complexes containing more than one T4Dam, but the

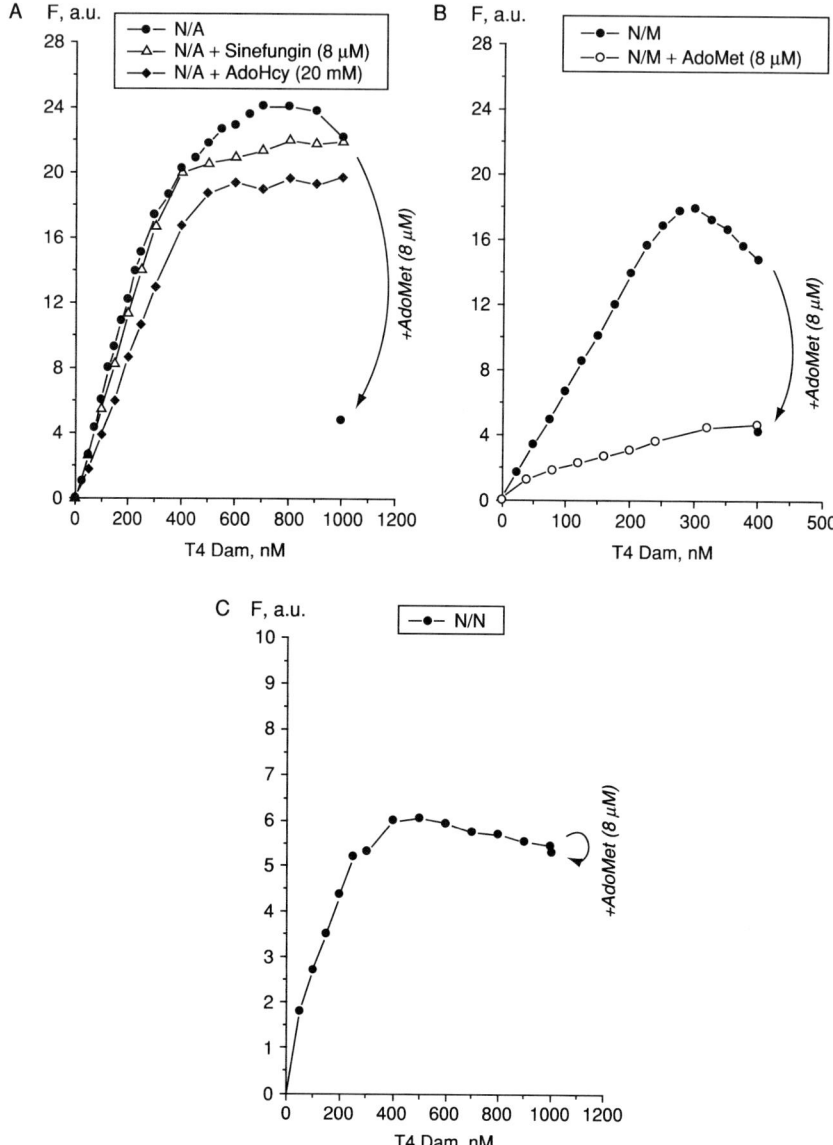

FIG. 2. Effect of added AdoMet on fluorescence analysis of T4Dam titration. Duplex N/A (A), duplex N/M (B), and duplex N/N (C) were at 200 nM; AdoMet was added to a final concentration of 8 µM at the last titration point. For duplex N/A (A), titrations are shown in the absence or presence of the reaction inhibitors AdoHcy (20 µM) or sinefungin (8 µM). The lower curve in (B) corresponds to a titration of duplex N/M in the presence of 8 µM AdoMet.

mechanism of this fluorescence quenching is not yet understood. Fluorescence of the N/M hemimethylated duplex also increased upon addition of T4Dam.

AdoHcy and analogue sinefungin are competitive inhibitors of AdoMet in transmethylation reactions. Presence of either AdoHcy or sinefungin did not dramatically reduce the fluorescence during T4Dam titration of duplex N/A or hemimethylated N/M duplex (Fig. 2A). In contrast to the methylation inhibitors, addition of substrate AdoMet led to a sharp reduction in the fluorescence intensity (Fig. 2B). Since duplex N/M is unable to serve as a methyl acceptor, the AdoMet-induced reduction in fluorescence occurs by a process independent of methylation per se.

We also carried out a T4Dam titration of a 20mer duplex having 2-AP substitutions in both strands (N/N duplex, Fig. 2C). It should be noted that in contrast to the asymmetric N/A and N/M duplexes, the doubly substituted N/N duplex possesses a high intrinsic fluorescence, 0.018 a.u./nM vs 0.0025 a.u./nM for N/A or N/M duplexes. This suggests that the two 2-AP residues in the N/N duplex were destabilized in comparison to the solitary N-residue in the other duplexes. For this reason, T4Dam interaction with the N/N duplex resulted in a reduced net increase in fluorescence, although the general shape of the titration curve was comparable to those for the N/A and N/M duplexes. However, when a saturating concentration of AdoMet was added at the titration end point with the N/N duplex, there was no decrease in fluorescence. This argues against either an alteration in the environment of the flipped-out base or a destabilization of the flipped-out 2-AP as the cause of the AdoMet-induced reduction in fluorescence, which should have reduced the fluorescence.

AdoMet binding to T4Dam is known to induce an allosteric alteration in T4Dam conformation, as demonstrated by tryptophan fluorescence quenching analysis (76). In this regard, methylation kinetics data (see Section III.B) indicate that the T4Dam–AdoMet complex was oriented approximately 50% of the time to the productive strand and, thus, bound randomly to the DNA duplex. Based on this, we proposed the following model. First, in the absence of AdoMet, T4Dam binds randomly to the N/A or N/M-duplex, and 2-AP is flipped out into the catalytic pocket and fluoresces in 50% of the duplexes. Second, addition of AdoMet induces an allosteric conformational change in T4Dam that causes the enzyme to favor and rapidly reorient to the "productive" strand. This reorientation results in flipping of the nonfluorescing A (or N^6-methylA) residue into the catalytic pocket, while the previously flipped 2-AP would no longer fluoresce when relocated in an intrahelical position. This rapid change in orientation between the nonproductive and productive strands could be achieved either by dissociation–reassociation or by reorientation without dissociation of the T4Dam–AdoMet complex from the DNA duplex. Since the recognition site within the N/N duplex is symmetrical, altering the binding orientation of T4Dam should have no influence on the fluorescence, since a 2-AP residue would be flipped out in every complex, consistent with the failure of added AdoMet to decrease the fluorescence of the T4Dam-(N/N) complex.

III. Kinetic Properties of T4Dam Methylation of Substrate Duplexes Containing Native or Altered Recognition Sites

A. Steady State Analysis

Originally, kinetic parameters of methyl transfer were determined using highly purified T4Dam and unmethylated, unglucosylated T4 gt^- dam^- DNA as the *in vitro* methyl acceptor (40, 63). However, due to the sequence complexity of a polymeric DNA substrate, there were numerous canonical GATC methylation sites, as well as potential noncanonical methylation sites. Therefore, we decided to utilize the same short defined synthetic duplex substrates (used to study binding) to investigate T4Dam methylation *in vitro* (51, 64).

Standard analyses of velocity dependence on substrate concentration were used to obtain catalytic rate and Michaelis constants for AdoMet and DNA, and to calculate specificity coefficient values. The K_m^{AdoMet} was calculated to be 490 nM with the 20mer specific duplex. Therefore, to determine kinetic parameters for different duplexes, AdoMet was used at a concentration of 2 μM. Adenine substitutions, A $\rightarrow$ P (purine) and A $\rightarrow$ N, were introduced to investigate the effect of lacking one target N6 amino group (and the loss of one Watson-Crick H-bond). The A $\rightarrow$ P and A $\rightarrow$ N substituted duplexes remained good substrates for methylation (in contrast to absence of an A), although eliminating the N6 amino group reduced the k_{cat} (Table I). Surprisingly, the A $\rightarrow$ P substitution also led to a reduction in the K_m for DNA. This may be related to the situation observed with EcoRI, where A $\rightarrow$ P substitution (GAPTTC instead of GAATTC) produced an association constant even higher than that for the canonical substrate (65), presumably because the substitution facilitated DNA distortion in the middle of the binary complex.

Previously, the importance of the G residue in the GATC-target site for both T4Dam binding and catalysis was observed (48, 64). In order to delineate the possible role in catalysis by each chemical substituent of G, we investigated the effect of replacing the imidazole N7 with a carbon atom (G $\rightarrow$ Z) or deleting the O6 keto group (G $\rightarrow$ N), as these are the most likely positions for major groove H-bonding with T4Dam (50, 66). The G $\rightarrow$ Z substitution was found to have a weak effect expressed only in slight increase in K_m value, which resulted in a 1.5-fold decrease of specificity coefficient compared to unmodified canonical duplex. This can be attributed to a single noncritical protein–DNA contact loss, weakly affecting the conversion step and overall course of reaction. In contrast, the G → N substitution greatly affected the kinetic parameters (25-fold reduction in specificity coefficient). This is almost comparable to the effect of deleting a G residue in one strand of recognition site (64). Thus, we conclude that T4Dam forms a critical contact with G at the O6 position.

In sharp contrast to their effect on binding, most defects in duplex structure dramatically reduced the rate of T4Dam methyl transfer; the highest relative methylation rates were observed only when a continuous GAT sequence was present on both duplex strands (64). For example, lack of an internucleotide phosphate between the central A and T (duplex 8) impaired methyl acceptor ability (4-fold increased K_m^{duplex} and about 3-fold lower k_{cat}), although the stability of the complex increased. While absence of the internucleotide phosphate between the A and G (duplex 9) residues had no effect on K_m, the k_{cat} value was decreased 10-fold, and complex stability was increased. In contrast, lack of an internucleotide phosphate between the C and T residues in the bottom strand (duplex 7) had little effect on the K_m^{duplex} and k_{cat} values relative to the 20mer specific duplex, although the stability of the complex between this duplex and T4Dam is dramatically reduced.

The kinetic parameters for T4Dam determined at 25 °C for short duplex substrates differed substantially from those obtained at 37 °C with unglucosylated, unmethylated polymeric T4 gt^- DNA as the substrate (40). The K_m values for ODNs were one to two orders of magnitude higher than with T4 gt^- DNA (calculated for molar concentrations of GATC sites). In addition, k_{cat} values were approximately one order of magnitude lower with synthetic duplex substrates. These differences are, to some extent, influenced by the temperature, but the nature of the substrate probably plays a greater role. For polymeric T4 gt^- DNA, there is a different type of catalytic turnover, i.e., the enzyme acts processively, moving to other sites along the DNA chain by linear diffusion without dissociation (67); this will be discussed further in Section III.D. Finally, comparison of ODN duplexes to polymeric DNA is also complicated by the fact that the latter contains noncanonical sequences that can be methylated by T4Dam. However, the relative k_{cat}/K_m values for the ODN duplexes (Table I) may be useful for gauging the influence of different defects in natural DNA on their methylation by T4Dam. Note also that, compared to the EcoDam MTase, T4Dam is more sensitive to distortion of their common target site (64).

By and large, the data in Table I reveal an absence of any correlation between K_d and K_m values with various duplexes. This discrepancy can not be explained by a simple Michaelian scheme. The most plausible explanation can be found within the framework of an irreversible scheme. At saturating AdoMet concentrations, one can consider the one-substrate reaction:

$$E + D \underset{k_{-1}}{\overset{k_1}{\leftrightarrow}} E \cdot D \overset{k_2}{\to} E \cdot mD \overset{k_3}{\to} E + mD \quad (A)$$

where mD is the methylated product from duplex D (substrate AdoMet and product AdoHcy are omitted here). At least in several cases, including T4Dam, the irreversibility of the DNA methylation reaction was shown (68–70). In steady state conditions, initial velocity of this reaction is expressed as:

$$[E]/V = (1/k_3) + (1/k_2) + (1/k_1 \cdot [D])(1 + k_{-1}/k_2) \quad (1)$$

where [E] = total enzyme concentration and [D] = free duplex concentration. This expression can be put in the standard Michaelian form with:

$$k_{cat} = k_2 \cdot k_3/(k_2 + k_3) \quad (2)$$

$$k_m = k_3 \cdot (k_{-1} + k_2)/(k_1 \cdot (k_2 + k_3)) \quad (3)$$

It was demonstrated for EcoRI MTase that the rate constant for conversion of the central complex (MTase-DNA-AdoMet) to product (MTase-methylated DNA-AdoHcy) is at least 300-fold larger than k_{cat} (71). Based on the similar burst kinetics, it is reasonable to suppose that there is a similar relationship for T4Dam, i.e., that the rate constant of the chemical conversion step (k_2) is larger than the rate constant of the central complex dissociation (k_{-1}). The latter assumption is true, at least for the complexes that are detected by the gel shift assay. With these two assumptions, we obtained the simplified expressions:

$$k_{cat} = k_3 \quad \text{and} \quad k_m = k_3/k_1 \quad (4)$$

As is evident from this, the K_m value does not depend on the dissociation rate constant, k_{-1}, in agreement with the observed absence of correlation between K_m and the apparent K_d (Table I). These kinetic examples show once more, that the old rule "better binding = better catalysis" has limited utility.

B. Pre-Steady State "Burst" Analysis

Addition of T4Dam to a mixture of radio-labeled AdoMet and unlabeled 20mer ODN duplex resulted in a "burst" of product (methylated duplex) formation, as shown in Fig. 3. This burst was followed by a constant rate of product formation, which reflected establishment of steady state conditions. The simplest interpretation of these data was that an initial rapid methylation was followed by a slower step, liberation of product(s) (AdoHcy and/or methylated duplex) from the enzyme. Thus, as in cases of the EcoRI, MvaI, PvuII, and CcrM MTases, the chemical step (catalysis of methyl group transfer) is not the rate-limiting step in the reaction (72–75). Bursts have also been registered for HhaI (76, 77), MspI (69), RsrI (61), EcoDam (78, 79), KpnI (80), and BamHI (81).

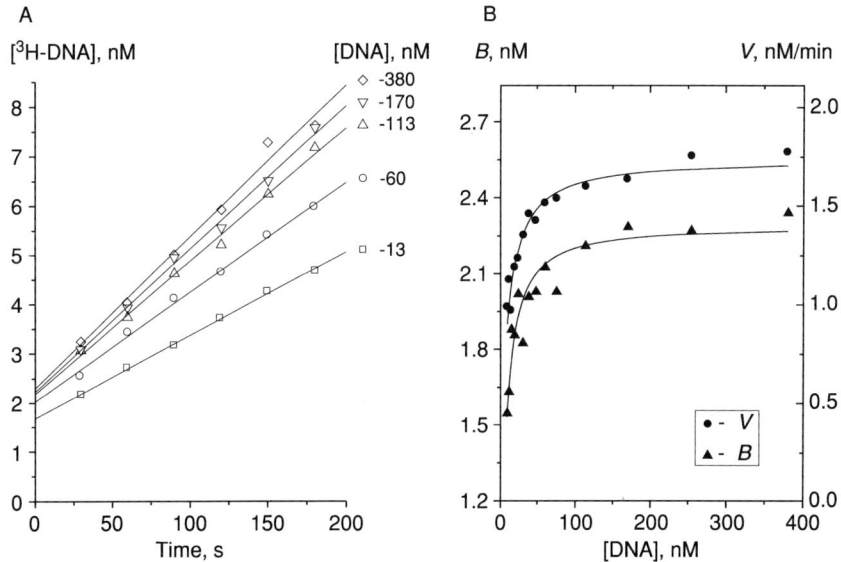

FIG. 3. Time course of T4Dam methylation of synthetic ODN duplexes as a function of DNA concentration (canonical duplex 6, see Table I) (A). T4Dam concentration was 2.25 nM (B). Initial reaction rates (V) and "bursts" (B) for duplex 6 methylation by T4Dam (2.25 nM) as a function of DNA concentration. Taken from (45) by permission of Oxford University Press.

The burst magnitudes, B, were defined as the y-axis intercepts resulting from extrapolation of the linear portion of the curve to zero time according to the equation $[^3H\text{-}DNA] = B + Vt$, where V values reflect catalytic rate constants. The dependencies of B and V on DNA concentration are presented in Fig. 3. Both have a similar character, and the V values are approximately proportional to the B magnitudes. If B values correspond to a concentration of the complex between T4Dam and product (radio-labeled duplex), then the V values may be represented as $V = B \cdot k_{cat}$.

To analyze this in more detail, the initial rapid methylation was studied using a rapid quench instrument (82). Formation of T4Dam product over time for native unmethylated and hemimethylated ODN duplexes is shown in Fig. 4. Assays were performed under initial velocity conditions of saturating substrate (Fig. 4A, B, C); the reactions were started by addition of substrate DNA to preformed T4Dam-AdoMet complexes (Fig. 4A, B) or by addition of AdoMet to preformed T4Dam-DNA complexes (Fig. 4C). Under the assay conditions, presence of saturating concentrations of AdoMet (8 μM, $K_m = 0.49$ μM) and DNA (1 μM, $K_m = 5.3$ to 12.9 nM (51, 64)), assured absence of any free form of the enzyme. Thus, the methylation reaction scheme (A) can be shortened to:

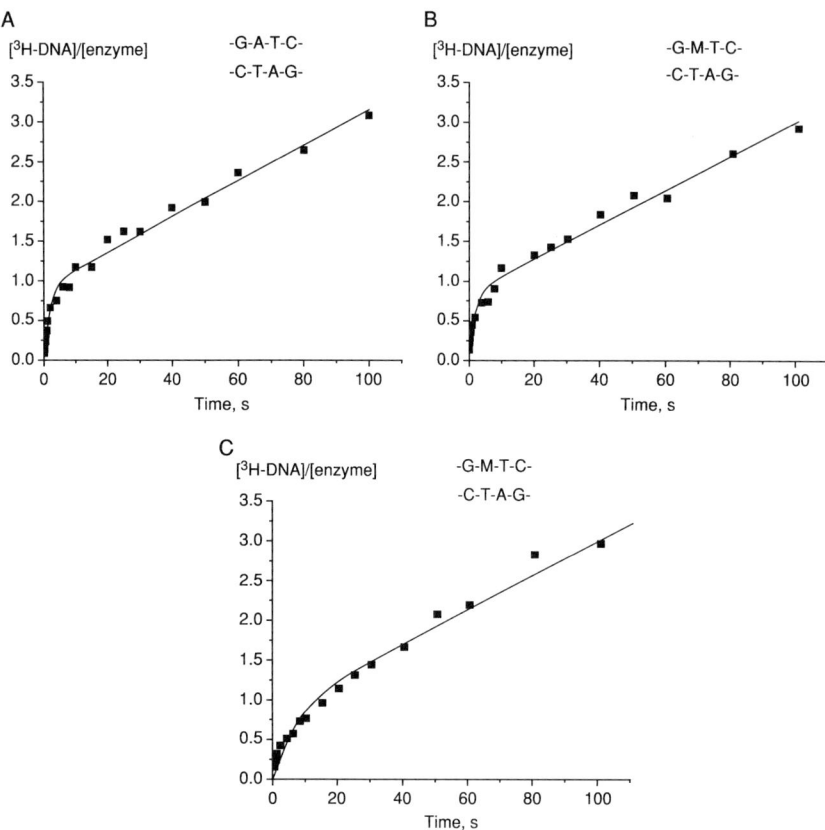

FIG. 4. Pre-steady state kinetics of the MTase using the preformed MTase–AdoMet complex (A, B) and the preformed MTase–DNA complex (C). The concentrations of T4Dam and 20mer duplexes were 0.158 μM and 1.0 μM, respectively; [^{3}H-CH$_3$] AdoMet was at 8 μM. The data were fit to Eq. (6). Taken from (82) by permission of Oxford University Press.

$$E \cdot D \underset{k_3}{\overset{k_2}{\leftrightarrow}} E \cdot mD \qquad (B)$$

where E is T4Dam with bound substrate AdoMet (or bound product AdoHcy), D is the substrate duplex, mD is the methylated product duplex, k_2 is the methylation rate constant, and k_3 is the rate constant for release of products mD and AdoHcy. According to this scheme, product accumulation is described by the equation:

$$[^3\text{H-DNA}]/[\text{enzyme}] = k_2^2 \cdot (1 - e^{-(k_2+k_3)\cdot t})/(k_2 + k_3)^2 + k_2 \cdot k_3 \cdot t/(k_2 + k_3) \tag{5}$$

While this equation fit our experimental data, a better fit was obtained with the equation in its more general form:

$$[^3\text{H-DNA}]/[\text{enzyme}] = B \cdot (1 - e^{-k_{\text{meth}}\cdot t}) + k_{\text{cat}} \cdot t \tag{6}$$

Here, B is the burst of product normalized to enzyme concentration, k_{meth} is the rate of methyl transfer, and k_{cat} is the steady-state reaction rate constant.

Fig. 4A shows T4Dam product formation over time with unmethylated DNA (duplex 1, Table II). Kinetic parameters derived from the data according to Eq. 6 are shown in Table II. The steady-state rate of product formation k_{cat} of 0.023 s^{-1} is in agreement with the value of 0.015 s^{-1} previously determined under initial velocity conditions (64). The value of k_{meth} (0.56 s^{-1}) was 24-fold higher than k_{cat} and a burst of product formation was observed. The burst value B of 0.92 methyl-groups transferred per enzyme molecule was close to the theoretical value of 1.0. The nonzero time course intercept indicates that catalysis was

TABLE II
PRE-STEADY STATE KINETIC PARAMETERS OF T4DAM INTERACTION WITH 20MER DUPLEXES CONTAINING A NATIVE OR MODIFIED RECOGNITION SITE AT LIMITING ENZYME CONCENTRATION[a]

Duplex no.	Recognition site[b]	Burst[c]	k_{meth},[c] s^{-1}	k_{cat},[c] s^{-1}
Canonical sites				
1	-G-A-T-C-	0.92	0.56	0.023
	-C-T-A-G-	(0.05)	(0.10)	(0.002)
1m	-G-M-T-C-	0.85	0.47	0.021
	-C-T-A-G-	(0.05)	(0.09)	(0.002)
1m[d]	-G-M-T-C-	0.84	0.14	0.021
	-C-T-A-G-	(0.07)	(0.04)	(0.002)
Substituted sites				
2	-G-A-T-C-	0.86	0.49	0.0051
	-C-A-T-G-	(0.05)	(0.09)	(0.0005)
3	-G-A-T-C-	0.90	0.67	0.0038
	-C-T-N-G-	(0.03)	(0.12)	(0.0005)

[a]Enzyme was pre-incubated with [^{3}H-CH$_3$]-AdoMet prior to the addition of duplex.
[b]N = 2-aminopurine; M = N^6-methyladenine.
[c]Values in parenthesis correspond to standard deviations.
[d]Enzyme was pre-incubated with duplex 1m before the addition of the [^{3}H-CH$_3$]-AdoMet.

limited by an event after methyl transfer. This is similar to results found for EcoRI (71), PvuII (74), HhaI (76), RsrI (61), and other DNA MTases.

Figure 4B shows the time course of product formation for hemimethylated DNA (duplex 1m, Table II). The derived kinetic parameters (Table II) were similar to those for the unmethylated substrate, with a k_{cat} of 0.021 s^{-1} and k_{meth} of 0.47 s^{-1}. If monomeric T4Dam MTase were to bind in a random orientation to hemimethylated duplex 1m, then a nonproductive complex should be formed in one-half of the binding events. However, the burst value of 0.85 was not consistent with this assumption. Under conditions of saturating substrates, the order of preincubation of T4Dam MTase with AdoMet or hemimethylated 20-mer duplex 1m had no influence on the burst value or k_{cat}. However, k_{meth} was approximately 3.5-fold lower when the enzyme was preincubated with duplex 1m than when it was preincubated with AdoMet (Fig. 4C; Table II). This suggests that in the first round of catalysis, the preformed T4Dam MTase–DNA complex is less efficient than the preformed enzyme–AdoMet complex. An alternative explanation would be that the preformed enzyme–DNA complex is not catalytically competent, must dissociate, bind AdoMet, and then rebind DNA, which could slow down productive complex formation.

Having a central double mismatch in the GATC sequence (duplex 2) had relatively minor effects on the parameters B and k_{meth}, but reduced k_{cat} by 4.5-fold (Fig. 5A, Table II). The double mismatch was expected to facilitate target A-flipping out of the DNA helix. However, since this defect only mildly decreased k_{meth}, A flipping is not the rate-limiting step in the methyl transfer reaction. It was shown that the EcoRI MTase has no specificity in its binding orientation, since the enzyme methylated only 50% of the hemimethylated-substrate duplexes under single turnover conditions (71). If the initial binding of T4Dam is random, then there are two alternative orientations (productive and nonproductive) of T4Dam on the asymmetric DNA duplex. Thus, a single binding event should have led to methylation of only 50% of the bound DNA duplexes. But, as seen from Table II, burst values are close to one. This indicated that for every binding event, an A residue was methylated. This requires that a T4Dam molecule oriented to the nonproductive strand be able to reorient to the productive strand without dissociation. Therefore, in order to modify all the A residues in the M/A or A/N asymmetric duplexes during a burst, the T4Dam–AdoMet complex must have either discriminated between the two strands when it initially bound (binding only to the A-containing productive strand), or it must have undergone rapid reorientation after binding in the nonproductive orientation. This suggests that if the initial binding is random, then there are two distinct pathways leading to a complete burst; the first pathway occurs when the enzyme binds to the productive strand, and

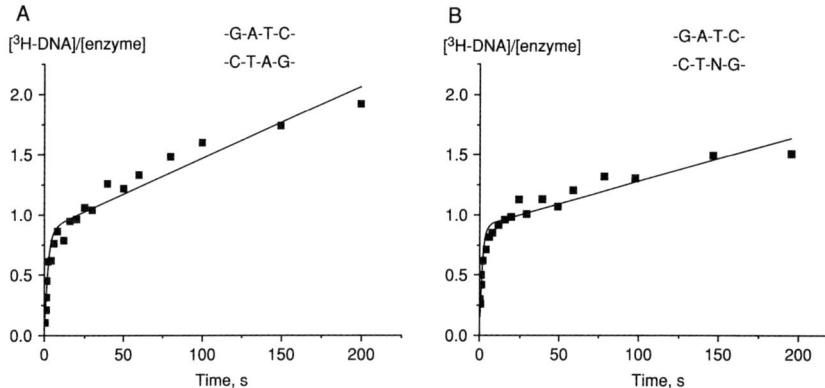

FIG. 5. Time course of T4Dam methylation of duplex having a double mismatch in the GATC sequence (A) and duplex N/A (B). T4Dam and duplex concentrations were 0.158 μM and 1.0 μM. T4Dam was pre-incubated with 8 μM [^{3}H-CH$_3$] AdoMet in both cases. The data were fit to Eq. (6). Taken from (82) by permission of Oxford University Press.

the second pathway occurs when the enzyme binds to the nonproductive strand and has to reorient. Thus, binding to the productive strand represents one step and binding to the nonproductive strand represents a second distinct intermediate step to a complete burst.

Although we have not presented direct experimental evidence that the T4Dam–AdoMet complex does not strand-discriminate during the initial binding (as opposed to binding randomly and undergoing reorientation), our data best fit the reorientation model. Thus, when a T4Dam–AdoMet complex collides with the M/A or A/N asymmetric duplexes under burst conditions ([enzyme] ≪ [duplex]), it binds at random. Methylation proceeds rapidly only with the T4Dam correctly oriented to the target A (productive strand). A slower, second phase of the burst ensues following reorienting of the T4Dam, which was initially oriented to the nonproductive M or N strand. Since burst conditions are normally found *in vivo* (a large excess of DNA substrate to enzyme), our model best describes how T4Dam normally interacts with its substrate. In this regard, it is interesting to note that, in the case of the RsrI MTase, the burst value for a hemimethylated 14mer ODN was also one (61). If the RsrI MTase bound randomly to the duplex and was unable to reorient when it bound to the nonproductive strand (containing M), a burst value of 0.5 should have been observed. However, since there was complete methylation of the hemimethylated DNA substrate, as with T4Dam (82), it appears that the RsrI MTase is capable of reorienting when it interacts with an asymmetrically modified recognition site. It would be interesting to determine whether other [N6-Ade] or [N4-Cyt] DNA MTases are capable of reorientation, and whether AdoMet plays a role in strand specificity for these enzymes.

In summary, the role of AdoMet in the methylation reaction of T4Dam goes beyond that of simply serving as a methyl donor. Thus, AdoMet-binding induces a conformational rearrangement in T4Dam (83). This not only results in an increase in the affinity of the enzyme to bind specific DNA containing its recognition site (48, 51), but it also results in an increased specificity for the strand containing the target base within the recognition site (45). According to our model, as a consequence of the AdoMet-induced conformational rearrangement and increased specificity, T4Dam–AdoMet is capable of undergoing a rapid reorientation to the productive strand in an asymmetrically modified recognition site. Such a model suggests that the T4Dam–AdoMet-DNA complex is not a static structure, but is in rapid dynamic equilibrium between two states of the complex, $E \cdot SD \leftrightarrow D \cdot SE$, which have different enzyme–AdoMet–DNA orientations. We have demonstrated this with the asymmetric N/A (Fig. 5B) and N/M duplexes, but the situation with the physiologically significant M/A remains to be determined. It should be mentioned, however, that we have observed a burst of approximately one using preformed T4Dam–AdoMet + M/A duplex, as well as with preformed T4Dam–M/A duplex + AdoMet (82) (Fig. 4). Biologically, it makes sense for the MTase to be able to reorient from a strand containing M to the one containing A without dissociating from the DNA, so as to increase the efficiency of methylation of hemimethylated DNA produced by replication *in vivo*.

C. Steady State Mechanism: Kinetic Evidence for a Catalytically Essential Conformational Change in the Ternary Complex

Elucidating the kinetic mechanism of the reactions catalyzed by DNA MTases still remains an important problem to investigate. Kinetic schemes have been proposed for HhaI (68, 76, 77, 84), MspI (69), human Dnmt1 (85) and murine Dnmt1 (86) [C5-cytosine] MTases, and for EcoRI (72, 87), EcaI (88), TaqI (89), EcoRV (62), EcoDam (79), KpnI (80), and Type III EcoP15I [N6-adenine] MTases (90). All of these enzymes exhibit a sequential bi–bi mechanism; however, they differ with respect to the order of substrate binding and rate-limiting step. For instance, whereas the rate of methyl group transfer is at least 300-fold faster than the rate of dissociation of the products for EcoRI (71), in contrast, the rate of methyl group transfer is the rate-limiting step in the TaqI methylation reaction (89). Furthermore, several different modes of MTase binding to substrates DNA (87, 91) and AdoMet or its analogs (84, 92) have been distinguished; in addition, changes in enzyme conformation (isomerization) associated with binding substrate DNA and/or AdoMet may